# Signal and Image Processing Techniques for Defense, Security, and Healthcare

B. Omkar Lakshmi Jagan
*Vignan's Institute of Information Technology, India*

Amrit Mukherjee
*University of South Bohemia, Czech Republic*

Thayyaba Khatoon Mohammed
*Malla Reddy University, India*

Vustikayala Sivakumar Reddy
*Malla Reddy University, India*

IGI Global
Scientific Publishing
Publishing Tomorrow's Research Today

| | |
|---|---|
| Vice President of Editorial | Melissa Wagner |
| Managing Editor of Acquisitions | Mikaela Felty |
| Managing Editor of Book Development | Jocelynn Hessler |
| Production Manager | Mike Brehm |
| Cover Design | Phillip Shickler |

Published in the United States of America by
IGI Global Scientific Publishing
701 East Chocolate Avenue
Hershey, PA, 17033, USA
Tel: 717-533-8845
Fax: 717-533-8661
Website: https://www.igi-global.com E-mail: cust@igi-global.com



Library of Congress Cataloging-in-Publication Data

Names: Omkar Lakshmi Jagan, B., 1990- editor. | Mukherjee, Amrit, 1988- editor. | Mohammed, Thayyaba, 1984- editor. | Reddy, Vustikayala, 1967- editor.
Title: Signal and image processing techniques for defense, security, and healthcare / edited by: B. Omkar Lakshmi Jagan, Amrit Mukherjee, Thayyaba Mohammed, Vustikayala Reddy.
Description: Hershey PA : IGI Global Scientific Publishing, [2025] | Includes bibliographical references. | Summary: "This book will attempt to give appropriate theoretical frameworks and the most recent empirical research findings in the field of signal and image processing techniques. The purpose of this book is to bring together researchers that create or develop sophisticated image processing techniques/systems, with a focus on defense, security and healthcare applications"-- Provided by publisher.
Identifiers: LCCN 2024053543 (print) | LCCN 2024053544 (ebook) | ISBN 9798369338407 (hardcover) | ISBN 9798369352038 (paperback) | ISBN 9798369338414 (ebook)
Subjects: LCSH: Signal processing--Digital techniques--Case studies. | Image processing--Digital techniques--Case studies.
Classification: LCC TK5102.9 .S5338 2025 (print) | LCC TK5102.9 (ebook) | DDC 621.382/2--dc23/eng/20250214
LC record available at https://lccn.loc.gov/2024053543
LC ebook record available at https://lccn.loc.gov/2024053544

British Cataloguing in Publication Data
A Cataloguing in Publication record for this book is available from the British Library.

# Table of Contents

# Detailed Table of Contents



A serious skeletal illness that lowers well-being and increases fracture risk, osteoporosis affects a large number of people worldwide. Early detection of osteoporosis in the knee is essential for successful treatment. This study presents deep learning, machine learning, and image processing techniques. It entails taking images of the knee joint using thermal imaging, x-rays, or other techniques, then processing them to improve quality and standardize format. The knee region is the target of methods such as region-of-interest (ROI) extraction. Numerous images are examined, encompassing geometric dimensions, bone mineral density, and textural features. Predictive models, such as decision trees, support vector machines (SVM), and convolutional neural networks (CNNs), classify knee joints as osteoporotic or healthy. The study also explores using an ensemble of multiple machine-learning models to improve system robustness and effectiveness.



Skin cancer, particularly dermo-cancer, is a critical health concern with rising incidences worldwide. Automated classification of dermo-cancer from skin images plays a pivotal role in early diagnosis and timely intervention. In this work, hybrid architecture that integrates inception and ResNet models to enhance feature extraction and facilitate hierarchical learning for improved dermo-cancer classification is explored. The inception module contributes to capturing multi-scale features, while the ResNet module addresses the challenges of vanishing gradients and aids in building a more robust and deeper neural network. The proposed hybrid architecture is trained on a comprehensive dataset, and experimental results demonstrate superior performance compared to individual models, achieving enhanced accuracy, sensitivity, and specificity. The approach automated dermo-cancer classification but also holds promise for other medical image tasks, showcasing the potential of hybrid architectures in medical image analysis.



Ovarian cancer is a common and reoccurring type of cancer. If detected early and treated properly, one can have a longer lifespan. We aim to provide robust alternative to the pre-existing system of manual diagnosis by automating the process using Deep Learning. In this paper we explore the three methods to classify histopathology images into its five corresponding types and outliers, possible cases of rare subtypes or healthy tissue misdiagnosed as cancerous. Previously, research made about detection and classification of other cancers have yielded significant results with several types of deep learning approaches, however the volume of data and annotations used for those is huge and difficult to obtain for all. Here, to make up for the lack of sufficient data, we have used transfer learning and fine tuning of pre-existing models trained for histopathological data.



The underwater world holds a wealth of secrets as it dominates three fourth of the earth surface. However, capturing underwater images with clear details is not an easy task. Light scattering, absorption, and limited visibility create a challenging environment for traditional cameras. The various phenomena that degrade underwater images and how it affect image clarity are examined. The image enhancement techniques, both traditional methods and cutting-edge deep learning approaches, which tackle issues like color correction, contrast enhancement, and haze removal, effectively restoring the true colors and details hidden beneath the water's surface are considered. Image enhancement plays a crucial role as a pre-processing step, paving the way for more accurate object detection. Various deep learning-based frameworks specifically designed to excel in these challenging conditions are explored in this chapter.



Emotions are vital in human communication, and EEG signals have become a potential method for detecting emotional states in real-time. Nevertheless, attaining a high level of precision in the identification of emotions continues to be a significant obstacle. Due to the emergence of deep learning techniques, there has been a growing desire to utilize these approaches for acquiring advanced characteristics in EEG emotion identification. This research makes a valuable contribution to the field of Human-Computer Interaction (HCI) medical applications. The purpose of this chapter is to offer a thorough examination of EEG emotion identification, specifically emphasizing the use of deep learning techniques. The initial focus of our study is a comprehensive exploration of the underlying principles and ideas presented in the existing literature. The primary objective of our research is to enhance the efficiency of emotion recognition models to provide more precise outcomes. This chapter offers insights into the most recent breakthroughs and potential areas for further inquiry.

This study introduces Dynamic SceneNet, a new Spatio-Temporal Interactive Network developed to tackle the issues of dynamic video foreground recognition. Proposed model uses sophisticated deep learning techniques to learn interactive spatiotemporal characteristics, resulting in a novel solution for accurate and adaptable foreground recognition in dynamic scenarios. Video foreground detection (VFD) is a crucial pre-processing step for accurate target tracking and recognition. Creating a reliable detection network is tough because to interference from shadows, changing backgrounds, and camera jitter. Convolution neural networks have shown to be reliable in several sectors due to their great feature extraction capabilities. This research proposes an interactive spatiotemporal feature learning network (ISFLN) for VFD.. CDnet2014, INO, and AICD datasets corroborate that the introduced ISFLN model surpasses state-of-the-art methodologies in video foreground detection, thanks to its prowess in feature enhancement, interactive multi-scale feature exploration, and deep learning techniques.



The utilization of brain imaging information holds promise for predicting the prognosis of diverse psychiatric disorders, including schizophrenia, anxiety, attention-deficit hyperactivity disorder, bipolar disorder, and depression. Researchers are displaying growing interest in harnessing Python applications like Nilearn, Nistats, and Nibabel packages to extract and analyze brain imaging data. The Python Nilearn package enables the decoding, analysis, and creation of predictive models using brain imaging data. With Nilearn, users can perform functional connectivity analysis, manipulate brain image volumes, and conduct advanced statistical research on brain images. Furthermore, Nilearn contributes to the development of standardized methods for extracting brain imaging information, addressing potential variations in imaging methodologies across studies. While there is limited literature available on the Nilearn package and similar tools used for similar purposes, it is crucial to promote Nilearn and similar software to facilitate comparisons between studiess.



Liver cancer is the sixth most common cancer. Liver cancer is frequently analyzed with Computed Tomography (CT) scan. Early diagnosis by CT could lead to high recovery rate, however going through all the CT slices for thousands or even millions of patients manually by professionals is hard, expensive, inefficient and prone to errors. The proposed method mainly focuses on precision of liver lesion segmentation and to improve performance of the CAD systems. The goal is to create a reliable Res-UNet model that can segment the liver, identify Regions of Interest (RoI) from nearby organs, and use that information to identify liver lesions using a different Res-UNet. The contributions of this chapter are identifying whether liver is healthy or affected by the disease (hepatocellular carcinoma), segmenting lesions from the extracted liver image using Res-UNet and to evaluate the accuracy of segmentation. The developed models have been tested using LiTS dataset. The accuracy of liver segmentation and lesion segmentation achieved using ResUNet are 0.9918 and 0.99, respectively.



This paper presents a novel watermarking method that utilizes the combined power of the Discrete Wavelet Transform (DWT), Discrete Cosine Transform (DCT),and Set Partitioning in Hierarchical Trees (SPIHT) algorithm. Our method embeds two distinct watermarks,into different sub-bands of host image.Initially, the cover image and Watermark1 undergo optimization using hybrid optimization techniques(Particle Swarm Optimization and the Butterfly Optimization). Subsequently, Watermark1

# Preface

The rapid advancements in signal and image processing have led to transformative applications in defense, security, and healthcare. As these fields continue to evolve, leveraging artificial intelligence, deep learning, and computational algorithms has become essential for addressing real-world challenges. This edited volume, *Signal and Image Processing Techniques for Defense, Security, and Healthcare*, presents a collection of innovative research contributions that demonstrate the power of image and signal processing techniques in solving complex problems across these critical domains. The chapters compiled in this book provide insights into novel methodologies, computational models, and cutting-edge frameworks that enhance the efficacy and accuracy of various applications.

This book is structured to highlight the significance of artificial intelligence-driven imaging and processing techniques. The included works explore various domains, such as early disease detection through medical imaging, deep learning models for security applications, and computational frameworks for efficient processing in high-stakes environments. By integrating interdisciplinary research and advancements, this book aims to bridge the gap between theoretical developments and their practical implementation. Each chapter presents comprehensive research backed by experimental results, providing readers with a well-rounded understanding of the subject matter.

## CHAPTER OVERVIEWS

### Chapter 1: A Perspective on Intelligent Imaging Techniques for Early Prediction of Knee Osteoarthritis

This chapter explores the application of deep learning, machine learning, and image processing techniques for early detection of knee osteoarthritis. It discusses various imaging modalities, including thermal imaging and X-rays, and details methodologies such as region-of-interest (ROI) extraction, feature analysis, and predictive modeling using decision trees, support vector machines (SVM), and convolutional neural networks (CNNs). The chapter also highlights the use of ensemble models to enhance system robustness and effectiveness.

## Chapter 2: Automatic Hybrid CNN-Based Skin Cancer Classification

Focusing on the automated classification of skin cancer, this chapter introduces a hybrid deep learning architecture integrating Inception and ResNet models for improved feature extraction. It explains how these models address challenges such as vanishing gradients and multi-scale feature representation. The chapter presents experimental results showcasing the superior accuracy, sensitivity, and specificity of the proposed approach, emphasizing its potential applications in medical image analysis.

## Chapter 3: Classification of Ovarian Cancer Subtypes Using Histopathology Images

This chapter presents a deep learning-based method for classifying ovarian cancer subtypes from histopathology images. It discusses the challenges associated with manual diagnosis and introduces transfer learning and fine-tuning techniques to compensate for limited data availability. The study evaluates the effectiveness of three classification methods, providing insights into improving automated histopathological analysis for ovarian cancer detection.

## Chapter 4: Computer-Aided Algorithms for Underwater Image Enhancement and Object Detection

Addressing the challenges of underwater imaging, this chapter examines light scattering, absorption, and visibility issues. It explores traditional and deep learning-based image enhancement techniques for color correction, contrast improvement, and haze removal. The discussion extends to object detection frameworks optimized for underwater conditions, emphasizing their role in maritime security and ecological studies.

## Chapter 5: Deep Learning Strategies for Analyzing EEG Signals in Emotion Recognition

This chapter investigates the use of deep learning models in detecting emotional states through EEG signal analysis. It provides a thorough review of existing literature, discusses fundamental principles, and presents strategies to improve recognition accuracy. The chapter emphasizes the role of deep learning in human-computer interaction (HCI) and medical applications, highlighting advancements in EEG-based emotion classification.

## Chapter 6: Dynamic SceneNet Spatio-Temporal Interactive Network for Dynamic Video Foreground Detection

This chapter introduces Dynamic SceneNet, a novel spatio-temporal interactive network designed for video foreground detection. It examines the challenges posed by changing backgrounds, shadows, and camera jitter. The proposed model leverages deep learning techniques to enhance interactive feature learning, outperforming state-of-the-art methodologies in video foreground recognition based on benchmark datasets.

## Chapter 7: Exploring Brain Imaging Analysis with Nilearn and Related Python Packages

This chapter delves into the use of Python-based tools such as Nilearn, Nistats, and Nibabel for brain imaging analysis. It outlines their functionalities in decoding, analyzing, and modeling brain imaging data for psychiatric disorder prognosis. The chapter emphasizes the standardization of brain imaging techniques and promotes the broader adoption of these tools in neuroscientific research.

## Chapter 8: Identification of Hepatocellular Carcinoma from CT Liver Images for CAD Systems

Focusing on liver cancer diagnosis, this chapter presents a Res-UNet-based segmentation model for detecting hepatocellular carcinoma from CT images. It highlights challenges in manual analysis and explores automated segmentation techniques to improve diagnostic accuracy. The chapter discusses the model's performance using the LiTS dataset and evaluates its precision in identifying liver lesions.

## Chapter 9: Secure and Robust Digital Image Watermarking Technique to Secure Medical Images

This chapter presents an advanced watermarking approach combining Discrete Wavelet Transform (DWT), Discrete Cosine Transform (DCT), and Set Partitioning in Hierarchical Trees (SPIHT) for securing medical images. It details encryption techniques and evaluates the robustness of the method against various attacks, demonstrating its high PSNR and NC values.

## Chapter 10: Secure Implementation of Image Processing Operations on CPU and GPU Architectures with Deep Learning Frameworks

This chapter explores the integration of deep learning frameworks with CPU and GPU architectures for secure image processing. It discusses performance trade-offs, security concerns, and methodological approaches, providing insights through statistical analysis and expert surveys. The chapter lays the groundwork for future research on secure and efficient deep learning-based image processing techniques.

The collaboration of esteemed researchers and scholars from institutions worldwide has made this book a valuable resource for academics, professionals, and students. By delving into the complexities of signal and image processing, the contributors offer cutting-edge solutions that pave the way for future innovations. It is our hope that this collection will inspire further research and development, fostering advancements that will shape the future of defense, security, and healthcare industries.

The chapters included in this book span a diverse range of topics, reflecting the multidisciplinary nature of image and signal processing. From intelligent imaging techniques for medical diagnosis to secure implementations of deep learning frameworks, the research presented here offers practical applications that address pressing global challenges. Techniques such as convolutional neural networks, hybrid architectures, and automated classification models are explored in detail, demonstrating their effectiveness in enhancing medical imaging, biometric security, and real-time surveillance.

Furthermore, the book delves into novel methodologies in underwater image enhancement, dynamic video foreground detection, and secure image watermarking. These research contributions illustrate the growing need for robust and adaptable computational approaches in diverse environments. By incorporating state-of-the-art techniques, the authors provide a forward-looking perspective on how artificial intelligence and deep learning continue to revolutionize signal and image processing applications.

We extend our gratitude to all the contributing authors, reviewers, and institutions whose dedication and expertise have made this book possible. Their collaborative efforts and commitment to research excellence have resulted in a comprehensive resource that will serve as a foundation for future advancements in this field. Additionally, we appreciate the support of our editorial team, whose meticulous work has ensured the quality and coherence of this volume. Their invaluable contributions have helped shape this book into a valuable reference for researchers and practitioners alike.

As we move forward in an era of rapid technological advancements, we believe that the insights shared in this book will play a significant role in shaping the future of signal and image processing. We hope that readers will find this collection both informative and inspiring, as it provides a glimpse into the potential of these technologies to transform industries and improve lives. We invite you to explore the research presented in this volume, and we look forward to witnessing the continued growth and innovation in this ever-evolving field.

**B. Omkar Lakshmi Jagan**
*Vignan's Institute of Information Technology, India*

**Amrit Mukherjee**
*University of South Bohemia, Czech Republic*

**Thayyaba Khatoon Mohammed**
*Malla Reddy University, India*

**Vustikayala Sivakumar Reddy**
*Malla Reddy University, India*

# Chapter 1
# A Perspective on Intelligent Imaging Techniques for Early Prediction of Knee Osteoarthritis

**Binit Roy**
https://orcid.org/0009-0006-1393-9633
*ICFAI University, Tripura, India*

**Debapriya Banik**
https://orcid.org/0000-0001-5031-1374
*ICFAI University, Tripura, India*

## ABSTRACT

*A serious skeletal illness that lowers well-being and increases fracture risk, osteoporosis affects a large number of people worldwide. Early detection of osteoporosis in the knee is essential for successful treatment. This study presents deep learning, machine learning, and image processing techniques. It entails taking images of the knee joint using thermal imaging, x-rays, or other techniques, then processing them to improve quality and standardize format. The knee region is the target of methods such as region-of-interest (ROI) extraction. Numerous images are examined, encompassing geometric dimensions, bone mineral density, and textural features. Predictive models, such as decision trees, support vector machines (SVM), and convolutional neural networks (CNNs), classify knee joints as osteoporotic or healthy. The study also explores using an ensemble of multiple machine-learning models to improve system robustness and effectiveness.*

DOI: 10.4018/979-8-3693-3840-7.ch001

# INTRODUCTION

Knee osteoarthritis, often known as KOA, is a progressive joint disorder characterized by the gradual deterioration of the protective cartilage that coats the ends of the knee bones. It is among the most prevalent types of arthritis, especially in the elderly. Osteoarthritis is an illness marked by pain, stiffness, and limited mobility due to protecting cartilage in the joint deteriorating over time. Although the precise origin of knee osteoarthritis is unknown, several variables are thought to be involved, such as ageing, genetics, obesity, overuse or damage to the joint, and misalignment of the joint. Bones may start to rub against one another as cartilage deteriorates; causing more injury and the development of bone spurs (Ondrésik, 2016).

The bones within the knee joint may experience friction, resulting in discomfort, rigidity, inflammation, and restricted mobility. Significance of identifying knee osteoarthritis early intervention and management aim to address and control the condition promptly. Pain relief strategies are implemented to alleviate discomfort. Preservation of joint function is prioritized to maintain optimal mobility. Preventing disability is a crucial objective to ensure long-term functionality. Enhancing quality of life is a central focus of the treatment approach. Customized treatment plans are developed to cater to individual needs. Preventing complications is a crucial aspect of the overall care.

The most prevalent type of arthritis that causes pain and mobility restrictions is knee osteoarthritis (KOA), a degenerative knee joint condition. Osteoarthritis of the knee (KOA) is a long-term musculoskeletal disease that impairs function and quality of life by causing persistent knee pain and incapacity. It is difficult to evaluate the degree and intensity of the pain that people experience since pain in KOA is subjective and idiosyncratic.

More than 250 million individuals globally suffer from knee osteoarthritis (OA), a chronic joint ailment that is the leading cause of disability. Knee OA causes the disease's majority (83% of the entire burden). Particularly in women over the age of 55, knee OA is more common (Ondrésik, 2016).

Timely detection of knee osteoarthritis (OA) is essential for mitigating disease progression, alleviating pain, and enhancing quality of life. Prompt recognition of osteoarthritis reduces the likelihood of surgical intervention by facilitating rapid management, which encompasses lifestyle modifications, physical therapy, and targeted treatments. However, the existing approaches for early diagnosis, such as x-rays and clinical evaluations, demonstrate limitations. They frequently exhibit insensitivity to nuanced variations, are subjective, and are susceptible to inconsistencies among observers. As a result, these methods often neglect pre-symptomatic or moderate cases. Thus, machine learning for osteoarthritis detection can analyze recognized patterns in imaging, biomarkers, and clinical data with enhanced precision.

This chapter offers many perspectives on knee osteoarthritis. Additionally, various sophisticated diagnostic methods for knee osteoporosis screening are thoroughly explored. This chapter further emphasizes the importance and difficulties associated with early osteoporosis prediction in the knee. This chapter aims to investigate several computer-aided intelligent imaging methods for knee osteoarthritis that may be practical and effective in clinical settings. In this manner, the chapter is structured: In Section 2, the incidence and current trends, symptoms, and risk factors are carefully covered. We discussed various diagnostic techniques in Section 3 for the early diagnosis of knee osteoarthritis. Section 4 discussed the necessity of early osteoarthritis prediction in the knee. We have outlined various obstacles to using intelligent imaging techniques to predict knee osteoarthritis early in Section 5. In Section 6, we have thoroughly examined many computer-aided innovative strategies by several research groups throughout the world for the early diagnosis of knee osteoarthritis. There are some noteworthy observations from the relevant literature and discussion of future directions in Section 7. In Section 8, the closing reflections for this chapter are reported.

*Figure 1. Healthy knee joint vs osteoarthritis knee joint (Ondrésik, 2016)*

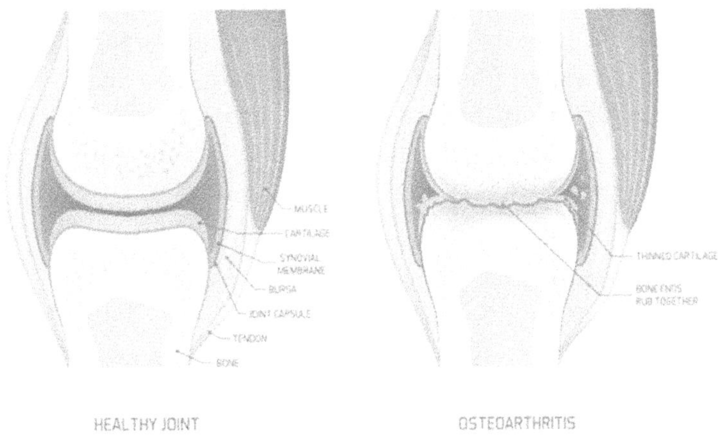


## 2. EPIDEMIOLOGY OF KNEE OSTEOARTHRITIS

The most prevalent joint condition, osteoarthritis, has risen to the fourth rank among all causes of disability globally in 2020. Eighty-three per cent of all cases of OA are related to the knee. There are few estimates on the prevalence of osteoarthritis in primary care settings. Conservative treatment for osteoarthritis (OA) aims to reduce pain, enhance joint function, and minimize joint deterioration. The three prior conventional therapies for knee OA are education about the condition, exercise recommendations, and encouragement of overweight or obese people to lose weight. Acetaminophen and non-steroidal anti-inflammatory medicines are two medications used in the pharmacological treatment of knee OA (NSAIDs). Multimorbidity, which includes conditions including obesity, diabetes mellitus, chronic obstructive pulmonary disease, and cardiovascular disorders, may influence decisions about medicines. Creating management plans tailored to multimorbidity will give pharmaceutical treatment specialized guidance (Spitaels, 2020). A clinical examination can establish the diagnosis, and X-rays can confirm it. Five grades of knee OA can be distinguished: Grade 0: The condition of the knee is "normal." Grade 1: Minimal growth of bone spurs without any pain or discomfort. Grade 2: This is the stage in which a person may initially exhibit symptoms. After a strenuous walking day, they will experience pain and notice increased joint stiffness. Even when the problem is only minor, X-rays will already show increased growth of the bone spur. Likely to stay in a healthy range in size is the cartilage. Grade 3: Mild OA. There will be more joint stiffness and frequent pain when moving, particularly in the morning and after extended periods of sitting. The distance between the bones is decreasing, and there is visible damage to the cartilage between them. Grade 4: The most severe OA stage is this one. There will be a significant drop in synovial fluid, nearly no cartilage left, and a reduction in the joint space between the bones. High levels of pain and discomfort when walking or moving the joint are typically linked with this stage.

*Figure 2. Healthy knee joint vs. osteoarthritis knee joint (Du, 2018)*

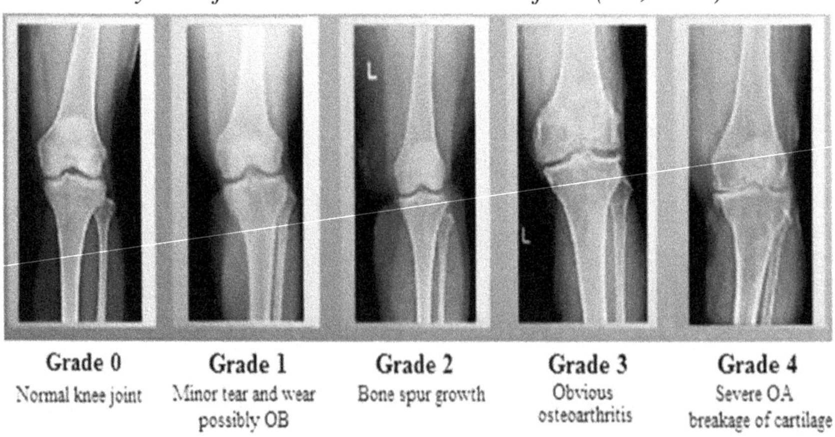

## 2.1 Current Trends in Knee Osteoarthritis

To determine the global prevalence and incidence of knee osteoarthritis (OA), a meta-analysis of 88 publications with 10,081,952 participants was performed. The combined worldwide majority of knee OA was determined to be 16.0% in people over the age of 15 and 22.9% in people over 40. Accordingly, it was projected that 654.1 million people (40 years of age and older) will have knee OA globally in 2020. According to estimates, the worldwide incidence of knee OA in people 20 years and older was 203 per 10,000 person-years. Further estimates put the number of people (20 years and older) with incident knee OA each year at 86.7 million worldwide in 2020. Age was a critical factor in the frequency and incidence of knee OA, which differed significantly throughout nations (Cui, 2020). For both genders, the corresponding ratios of prevalence and incidence were 1.69 and 1.39.

*Figure 3. Healthy knee joint vs. osteoarthritis knee joint (Cui, 2020)*

*Figure 4. Healthy knee joint vs. osteoarthritis knee joint (Cui, 2020)*

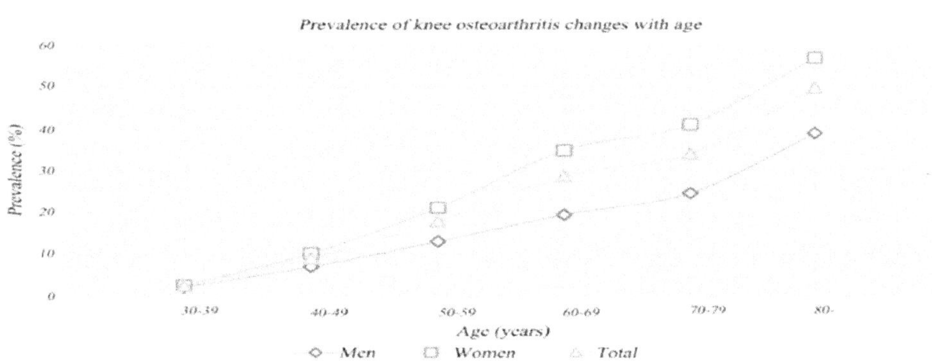

## 2.2. Etiology of Knee Osteoarthritis

Depending on its etiology, knee OA is categorized as primary or secondary. The cause of primary knee OA and articular cartilage degradation are unknown. This is commonly understood to be age-related deterioration in addition to normal wear and tear. The cause of articular cartilage degradation that leads to secondary knee OA is known reasons that could lead to secondary knee OA:- i. Hyper mobility or instability of the joints: An excessive range of motion in the joints is referred to as hyper mobility or instability of the joints. This condition frequently increases flexibility but raises the risk of pain and damage. ii. Metabolic disorders, like rickets: Poor bone development brought on by low vitamin D levels in metabolic disorders like rickets results in weak, brittle, or malformed bones. iii. Congenital disabilities: Congenital impairments are conditions that affect a person's physical or mental skills from birth as a result of circumstances that occurred during foetal development. iv. Joint malpositioning, such as valgus/varus posture: An irregular alignment that affects the bones or joints is known as joint malpositioning, and it includes postures such as valgus (outward) or varus (inward). v. Past joint damage, such as a fracture along the articular surface (tibial plateau fracture): The articulating surface of the joint is impacted by prior joint trauma, such as a tibial plateau fracture, which may result in long-term problems with stability and mobility. vi. Loss of movement and immobility: The afflicted joints become stiff, weak, and have a reduced range of motion due to immobility and loss of movement. vii. Past family members: Due to standard genetic variables or inherited features that affect joint health and osteoarthritis susceptibility, a familial history of knee osteoarthritis may increase the risk that the condition would develop in subsequent generations.

# 3. MEDICAL DIAGNOSTIC PROCEDURES FOR SCREENING OF KNEE OSTEOARTHRITIS

Medical diagnostic techniques such as X-ray, MRI, CT scan, and infrared imaging are essential for screening knee osteoarthritis. X-rays disclose osseous abnormalities and constriction of joint spaces. MRI offers intricate images of soft tissues, identifying cartilage deterioration and inflammation. CT scans provide three-dimensional imaging for accurate examination of joint structures. Infrared imaging identifies temperature discrepancies, emphasizing inflammation and alterations in circulation, facilitating non-invasive early diagnosis (Piccolo, 2023). Since X-rays and CT scans mainly identify extensive bone loss and structural damage, they are not very sensitive when it comes to early-stage osteoporosis of the knee. Magnetic resonance imaging (MRI) is best for assessing soft tissues, whereas infrared imaging is still in its experimental stages when it comes to evaluating bone health; both methods are costly and impractical for routine screening.

## X-Ray

X-rays can detect knee osteoarthritis by looking for changes in bone density, osteophyte production, and joint space narrowing. Radiologists evaluate these symptoms to help with diagnosis and therapy planning by identifying degenerative changes in the joint that are suggestive of osteoarthritis. X-rays are helpful in the diagnosis of osteoarthritis in the knee because they can clearly show bone structures, such as bone spurs and narrowing of the joint space, allowing for accessible and affordable diagnostic evaluations. In addition to helping with surgical planning when needed, they provide rapid imaging with little radiation exposure, which speeds up diagnosis, treatment planning, and long-term tracking of the course of the disease. Radiography offers quick, easy, and educational imaging for medical practitioners, making X-rays an invaluable tool in treating osteoarthritis in the knee.

*Figure 5. X-ray of healthy knee joint vs. osteoarthritis knee joint (Piccolo, 2023)*

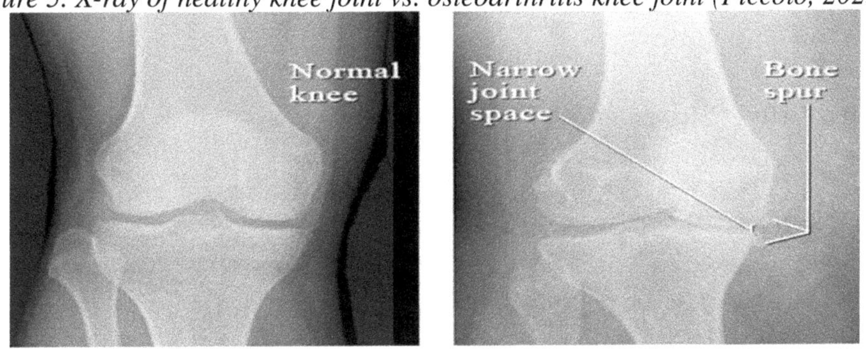


## MRI

Analyzing cartilage thickness, locating bone marrow lesions, and determining whether meniscus tears or degeneration are present are all necessary steps in detecting osteoarthritis in the knee using magnetic resonance imaging (MRI). By allowing for a thorough assessment of the course of osteoarthritis and providing guidance for treatment choices, these MRI results assist in illustrating structural alterations inside the knee joint.

*Figure 6. MRI of osteoarthritis knee joint (Piccolo, 2023)*

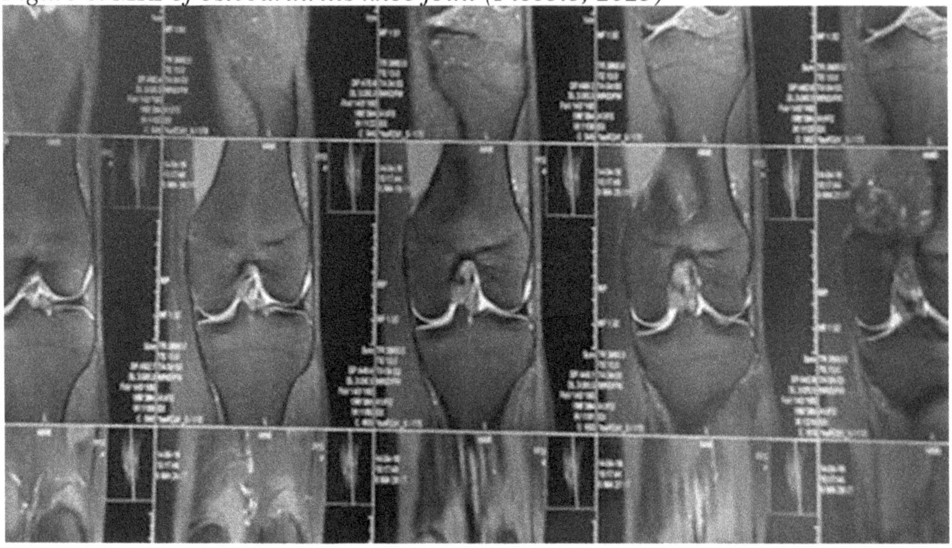

## Infrared Thermography

One really neat way to identify osteoarthritis in the knee is via infrared thermography. It functions by absorbing the infrared light that the body's surface emits. The affected knee typically has a slightly higher temperature because osteoarthritis is linked to inflammation and increased blood flow. Infrared imaging techniques offer several advantages for knee osteoarthritis (OA) detection: Non-invasive: Infrared imaging is non-invasive, meaning it doesn't require inserting instruments or needles into the body. This reduces patient discomfort and eliminates the risk of infection. Painless: Unlike other imaging techniques, such as X-rays or MRI scans, infrared imaging is painless. This makes it particularly suitable for patients who may experience discomfort during other imaging procedures. No radiation exposure: Unlike X-rays and CT scans, which use ionizing radiation, infrared imaging does not expose patients to harmful radiation. This makes it a safer option, especially for repeated imaging studies over time. Cost-effective: Compared to other imaging modalities like MRI, infrared imaging equipment is comparatively less expensive. This increases their accessibility in healthcare environments, particularly in places with little resources. Versatile and portable: A few infrared imaging tools are portable and suitable for usage in various therapeutic contexts, such as sports medicine clinics, primary care offices, and rehabilitation facilities. This adaptability makes it possible to quickly and easily monitor knee OA in various patient populations.

*Figure 7. Infrared image of osteoarthritis knee joint (Piccolo, 2023)*

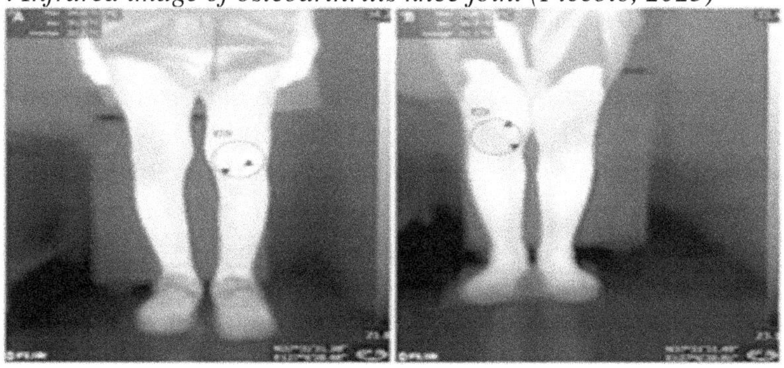

## CT Scan

CT is one medical imaging method that has significantly influenced evaluations and conclusions regarding health. This image represents a typical multimodal medical fusion. Like an X-ray, a scan creates cross-sectional images of a specific body part. For example, a CT scan of the knee can be used by doctors to look at injuries or illnesses on the knee. A CT scanner circles the body as images are sent to a screen. The computer uses these photos point by-point to illustrate its points of view. This enables experts in their field of expertise and training to view the bones, muscles, ligaments, tendons, and lungs that comprise an individual's leg. Similarly, a CAT scan is another name for a CT scan. The results are evaluated in a designated casualty test centre or a laboratory.

*Figure 8. CT scan of osteoarthritis knee joint (Piccolo, 2023)*

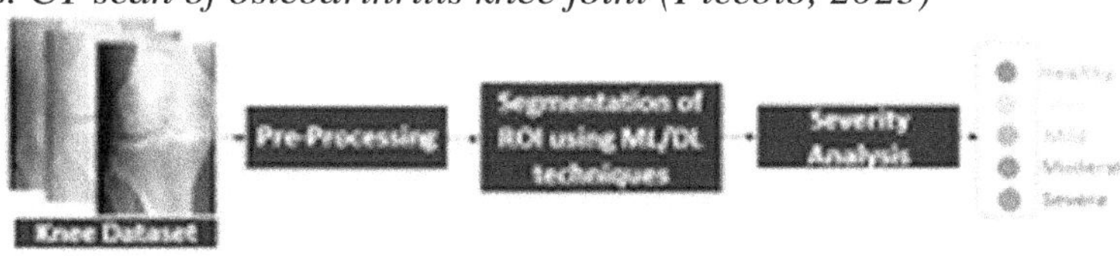

## 4. NECESSITY FOR EARLY PREDICTION OF KNEE OSTEOARTHRITIS

Because significant metabolic derangements in cartilage occur before the onset of superficial fibrillation, early identification of knee osteoarthritis (OA) is essential. A normal macroscopic appearance does not necessarily imply the absence of pathology. Conventional methods of assessing OA rely on traits after the degenerative process, which might not be sufficient to detect the illness early on. Early diagnosis enables the application of therapeutic approaches that may slow the progression of the disease, such as periarticular osteotomy and Femoroacetabular impingement (FAI) treatment. Early attention to mechanical issues, including joint morphology or alignment, can stop or reduce the advancement of the illness. Preventive detection also enables planning clinical trials for innovative therapies and identifying patients who may benefit from joint-preserving operations (Pollard, 2008).

## 5. OBSTACLES TO EARLY DETECTION OF KNEE OSTEOPOROSIS

Identifying knee osteoarthritis in its initial phases is challenging because traditional diagnostic methods for the condition focus on signs present in the disease's late pathological stages. These methods rely on an inaccurate indirect assessment of the condition of the articular cartilage. Only the advanced erosive stage of osteoarthritis exhibits joint space narrowing, a common sign of disease progression. More tactful methods are needed to monitor early illness and evaluate cartilage morphology. Alternative imaging methods, such as CT arthrography, might offer a more thorough evaluation of cartilage and spot early structural abnormalities (Rahmati, 2016). Loss of articular cartilage is a significant alteration in osteoarthritis, but secondary changes in the joint include bone remodeling and synovial changes. Genetics and biomarker assessment are being investigated as potential early detection strategies; however, more research is needed to evaluate their effectiveness.

## 6. EXAMINATION OF SOPHISTICATED MEDICAL IMAGING METHODS FOR EARLY PREDICTION OF KNEE OSTEOARTHRITIS

Regarding the early diagnosis of osteoarthritis in the knee, this section examines the steps in the advanced medical imaging treatments suggested in the last several years (Zebari, 2022). Using machine learning (ML) techniques to properly establish the region of interest by drawing out features and then training them with classifiers such as support vector machines (SVM), artificial neural networks (ANN), and so on.

*Figure 9. Steps for early prediction of KOA using intelligent medical imaging techniques*

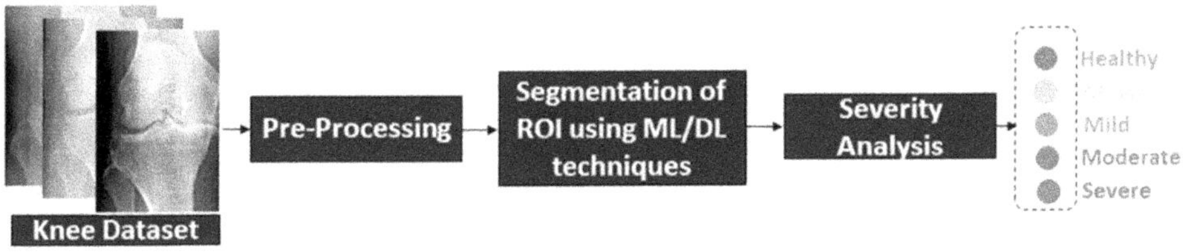


## 6.1 Collection of Public/Private Knee Datasets

Datasets give the study the fundamental support it needs by offering the variety of information required for an in-depth analysis. Researchers can collaborate to enhance knee health through greater transparency, reproducibility, and cooperation by using publicly available datasets. On the other hand, private databases make it possible to delve deeply into niche topics and do studies that would be impossible with publicly available data. Knee image dataset refers to a set of digital images that capture various aspects of the knee joint. These images can be obtained through medical imaging techniques such as X-rays, Magnetic Resonance Imaging (MRI), or thermography scans. Table 1 emphasizes the dataset's crucial importance in detecting knee osteoarthritis. Medical image analysis using machine learning methods, such as convolution neural networks (CNNs), highly depend on the training dataset's quantity, diversity, and quality. A well annotated and extensive dataset greatly enhances AI models' performance and generalization abilities.

*Table 1. Summary of existing knee datasets*

| Dataset | No. of Samples | Acquisition Technique | Size | Availability |
|---|---|---|---|---|
| Osteoarthritis Initiative(OAI) database (National Institute of Health) | contains large no. of X-ray, MRI, thermal images | X-ray, MRI, thermal camera | NA | On Request |
| OSTHERSIT (Sethu Institute of Technology) | 9 men and 21 women | Thermal Camera | 320x240 | On Request |
| Knee Osteoarthritis Dataset with Severity Grading (Kaggle) | 1526 | X-Ray | NA | Open Access |
| IR(Infrared) Knee Joint Dataset (IR(Infrared) Knee Joint Dataset) | 200 | Thermal Camera | NA | On Request |

## 6.2 Preprocessing

Initial datasets are frequently furnished with labels and tags for classification during the knee osteoarthritis detection preprocessing phase. These labels and tags are vital components that dataset suppliers add from the beginning. While the software can merge these components without problems, the imaging equipment is responsible for introducing undesired noise during photographing. It is vital to systematically eliminate extraneous labels, tags, and data to streamline the dataset and conduct targeted examinations. This is done to guarantee the accuracy of the following analyses. In addition, because the background that is taken alongside the knee photos might provide extra features, one of the essential steps entails removing this background to isolate the particular knee area. As a result of this thorough approach, not only does it improve the accuracy of subsequent studies by reducing noise, but it also optimizes computing efficiency, therefore laying the framework for efficient computer-aided techniques in knee osteoarthritis identification.

## 6.3 Segmentation of ROI Using ML/DL Techniques

State-of-the-art machine learning and deep learning techniques are used in knee osteoarthritis research to segment the ROI. Convolution neural networks, such as the U-Net and SegNet networks, are examples of networks that are able to accurately identify diseased regions in medical photographs. For the purpose of refining the segmentation, bigger datasets, strong training methods, and post-processing procedures are utilized. By making earlier and more accurate diagnoses, which are enabled by the constructed models, it is feasible to provide improvements in patient care. These models give clinicians important information about the body parts impacted by osteoarthritis. Knee joint Detection: Machine learning-based knee osteoarthritis detection relies heavily on identifying the knee joint in medical images. This detection is usually a component of a more extensive process that includes feature extraction, classification, and image preparation. In this case, the following is a summary of how to go about knee joint detection:

a. **Anatomical Landmarks**: To locate the knee joint, use anatomical landmarks or areas of interest (ROI). These markers may consist of the patella, tibia, and femur.
b. **Thresholding and Segmentation:** To isolate the region of the knee joint, use image thresholding and segmentation techniques. Using methods like Otsu's or region-growing algorithms can be one way to do this.
c. **Deep Learning:** Train a convolution neural network (CNN) for semantic segmentation using deep learning. For this, architectures such as U-Net and Mask R-CNN can be employed.
d. **Feature Selection:** Reducing computational complexity and noise, feature selection improves model accuracy in knee osteoporosis prediction by concentrating on the most important features. Better model performance on unseen data and the prevention of over fitting both contribute to improved generalization. Examples of feature selection approaches used in the prediction of knee osteoarthritis include recursive feature reduction, filter methods like chi-square and correlation analysis, and embedding methods like Lasso regression. In order to build reliable prediction models, these methods aid in determining which features are most important.
e. **Feature Extraction**: From the segmented knee joint region, extract pertinent features. Shape descriptors, texture features, and intensity histograms are some of these features. If your dataset is large enough to handle it, use deep learning-based feature extractors such as pre-trained CNNs. Computer Learning Framework: Teach a machine learning model—such as a deep neural network, random forest, or support vector machine (SVM)—to distinguish between osteoarthritis and normal knee joint images. Provide your classifier with the features that were extracted in the preceding step.
f. **Assessment:** Using relevant measures, including accuracy, precision, recall, F1-score, and ROC curves, evaluate the efficacy of your osteoarthritis classification and knee joint detection model.
g. **Adjustment and Verification:** Validate your model there to ensure that it performs effectively over various datasets. Make any necessary adjustments to your model based on the validation results.
h. **Utilization:** Once your model works properly, you can utilize it in a clinical setting to assist in detecting and diagnosing knee osteoarthritis.

i. **Model Interpretability:** Depending on the classifier being used, think of ways to display and comprehend the judgments made by the model. This is critical to understanding the classification's foundation.
j. **Deployment (if applicable):** If the model performs well enough, it might be used in a clinical setting to help diagnose osteoarthritis in the knee.
k. **Maintenance and Monitoring:** Retrain the model whenever fresh data becomes available and keep a close eye on its performance. It's critical to maintain the model current because medical datasets can change over time.
l. **Ethical Considerations:** Ensure your model respects moral principles and doesn't support prejudice or discrimination.
m. **Regulatory Compliance:** Before using the model in a clinical or medical setting, we must obtain the necessary permissions and abide by all relevant national, international, and local laws.

## 6.4 Severity Analysis

An extensive investigation of the extent of joint deterioration is required to assign a severity score to knee osteoarthritis in a range from 0 to 4. The grading system, which runs from a knee with no evident indications of abnormality (Grade 0) to a knee with considerable deformation of the bone and severe knee osteoarthritis (Grade 4), is a tool that supports the process of making decisions regarding medical care. Within the context of the progression of the disease, the grading process enables medical professionals to tailor treatments to the individual patient. It is essential to thoroughly assess the severity of knee osteoarthritis to design individualized treatment regimens and improve the general results for patients.

*Table 2. Summarized state-of-the-art detection and segmentation methods*

| **Author, Year** | **Database** | **Methods** | **Evaluation metrics** | **Remark** |
|---|---|---|---|---|
| (Du, 2018) | OAI | ANN | AUC: 903 | Missing information about computational time for the training and testing process using the ANN. |
| (Du, 2018) | Osteoarthrit Initiative (OAI) | A NN, Random Forest (RF), Naive Bayes(NB) | **ANN:**AUC: 0.761, F1 Score: 0.714 **RF:** AUC: 0.785, F1 Score: 0.743 **NB:**AUC: 0.695, F1 Score: 0.796 | Missing potential limitations or biases in the OAI database, which could affect the generalizability of findings. |
| (Brahim, 2019) | OAI | Random For-est (RF), Naive Bayes (NB) | ACC: 82.98% | Lack of calibration and consideration of soft tissue impact, exclusion of age and gender factors, and limitations of X-ray imaging for early disease detection. |
| (Moustakidis, 2019) | OAI | Random Forest (RF), Naive Bayes(NB) | Acc: 80.74% | Explores the possible use of quantum computing in the classification of osteoarthritis, but no actual implementation or assessment of this method is given. |
| (Jamshidi, 2020) | X-ray and magnetic resonance imaging | Logistic re gression | AUC: 0.80 | Used automatic machine learning to select best classification model. It introduce bias and limitations in the model selection process. |
| (Kwon, 2020) | Gait data and Radiographic image | Support vector machine (SVM) | sensitivity: 0.70, preci-sion: 0.76, F1-score: 0.71 | Used automatic machine learning to select best classification model. It introduce bias and limitations in the model selection process. |

continued on following page

*Table 2. Continued*

| Author, Year | Database | Methods | Evaluation metrics | Remark |
|---|---|---|---|---|
| (Wang, 2021) | OAI | Combination of YOLO Model | Acc: 95.57% | The use of deep learning models, such as YOLO and the visual transformer, may necessitate a large amount of computational time and re- sources for training. |
| (Bonakdari, 2021) | OAI | SVM | NA | Does not provide information on potential limitations or drawbacks of the algorithm. |
| (Olsson, 2021) | 6,103 knee radio- graphs from Danderyd University Hospital (2002-16) | CNN, Resnet | Mean AUC: 0.92 | There was a lack of testing of several deep learning architectures because the study's evaluation of a particular CNN of the ResNet architecture was its primary focus. |
| (Mahum, 2021) | Knee Os teoarthritis Severity Grading dataset (Mendeley dataset IV) | CNN, LBP, HOG | Acc: 97.14% | Hybrid feature extraction approach using CNN, LBP, and HOG may introduce complexity and computational overhead, potentially affecting the efficiency and scalability of the method. |
| (Pongsakonpruttikul, 2022) | OAI | YOLO version: 3 | Acc: 85% | Sample size used in the study was relatively small, which may have affected the over- all performance. |
| (Cueva, 2022) | Knee X-ray images | CNN, Resnet- 34 | Acc: 0.9,sensitivity: 0.885 and specificity: 0.917 | Dataset used for training the model had class imbalance issues, which can negatively impact the performance of the model. |

continued on following page

*Table 2. Continued*

| Author, Year | Database | Methods | Evaluation metrics | Remark |
|---|---|---|---|---|
| (Kanthavel, 2022) | OAI | Reinforcement learning (RL) | NA | Does not discuss any potential limitations or challenges associated with using deep learning and reinforcement learning techniques for medical image processing and diagnosis. |
| (El-Ghany, 2023) | OAI | DenseNet169 deep learning technique | Acc: 98% | Proposed model may vary when applied to different datasets or populations, as the evaluation was conducted on the OAI dataset. |
| (Tariq, 2023) | OAI | Transfer learning, ResNet-34, VGG-19, DenseNet 121, and DenseNet 161 | NA | Due to the processing needs and lengthy nature of the models, the deep learning-based technique might not be appropriate for real-time applications. |

## 6.3 Performance Metrics

In this section, we offer a succinct summary of the performance measures used to evaluate our classifiers before deploying machine learning models, it is crucial to assess their performance. Conventionally, we evaluate classifiers using classification accuracy and F1 scores. Classification accuracy is defined as the proportion of valid predictions to the total number of samples in the dataset (Equation-I). To derive relevant conclusions about the model based on its classification accuracy, it is crucial to ensure that the dataset is balanced. This is because achieving high classification accuracy on an unbalanced dataset may be due to a greater proportion of accurate predictions in the class that has a larger number of samples. Classes with a smaller number of samples have a lower impact on the final accuracy. The F1 score is an alternative method for assessing the performance of a model. We calculate the F1 score by taking the harmonic mean of precision and recall ratings (Equation-IV). We obtain precision (Equation-III) by dividing the number of accurately identified genuine positives by the total number of samples classified as positive. The model's high precision indicates its dependable ability to accurately categories samples as positive. Recall, however, refers to the proportion of true positives in relation to the total number of samples in the collection (Equation-III). A high recall indicates the model's ability to accurately identify positive samples as positive.

$$Accuracy = \frac{TP + TN}{FP + FN + TP + TN} \text{ (I)}$$

$$Precision = \frac{TP}{TP + FP} \text{ (II)}$$

$$Recall = \frac{TP}{TP + FN} \text{ (III)}$$

$$F1\ Score = \frac{2 \times Precision \times Recall}{Precision + Recall} \text{ (IV)}$$

## 7. DISCUSSION AND FUTURE DIRECTION

This research undertaking is both interesting and relevant because it seeks to use deep learning and machine learning to diagnose knee osteoarthritis. If we want to make better diagnoses of knee osteoarthritis, we might look at using cutting-edge machine learning techniques like deep learning. Incorporating multimodal data, which integrates clinical and patient history with imaging data (X-rays, infrared pictures, and magnetic resonance imaging), is one possible strategy to examine. Creating a model that can predict how osteoarthritis will proceed while also providing personalized treatment options would be a huge leap forward in the healing process. The use of explicable AI approaches has the potential to increase the level of trust that medical practitioners have in the model. When compared to more conventional approaches such as X-ray, MRI, and CT scans, machine learning algorithms provide superior early detection accuracy when it comes to knee osteoarthritis data analysis. Their ability to handle massive datasets enables predictive modeling and individualized risk evaluations. Furthermore, the addition of more data to machine learning algorithms enhances their ability to produce automatic outcomes. We can learn more about the model's decision-making process by using these strategies. Having an accessible network of medical professionals is crucial during the full development process. This makes it considerably easier to ensure that the AI models align with clinical practice needs and seamlessly integrate into the healthcare system's operations. Prior to implementing AI in healthcare, it is essential to address ethical concerns and guarantee the security of patient data. In light of upcoming advancements, it is critical to prioritize the establishment of moral norms and provide adequate data security.

## 8. CONCLUSION

It is feasible that the knee osteoporosis prediction approach that combines deep learning and machine learning might be of significant service to future medical research. This is something that is now being investigated. These technologies have the potential to completely transform the practice of diagnosing osteoporosis at an early stage and with greater precision. In addition to this, they have the capacity to facilitate prompt intervention and individualized treatment regimens, which is something that has never been successfully accomplished previously. The utilization of huge data sets enables these algorithms to achieve a higher level of diagnostic precision, which ultimately leads to the development of healthcare solutions that are not only more efficient but also economically more viable. For this potential to be fully realized, however, it will require collaboration between data scientists, developers of information technology, and medical professionals who are experts in their respective fields. Furthermore, ethical considerations, data privacy, and algorithm openness are all obstacles that need to be addressed in order to ensure widespread acceptance and implementation. Therefore, it is imperative that these issues be addressed. Deep learning and machine learning offer hope for improving knee osteoporosis patients' quality of life when combined with medical diagnostics. Moreover, this possibility presents itself due to the continuous progress in research and technology.

## REFERENCES


Bonakdari, H., Jamshidi, A., Pelletier, J.-P., Abram, F., Tardif, G., & Martel-Pelletier, J. (2021). A warning machine learning algorithm for early knee osteo-arthritis structural progressor patient screening. *Sage (Atlanta, Ga.)*, *13*. DOI: 10.1177/1759720X21993254 PMID: 33747150

Brahim, A., Jennane, R., Riad, R., Janvier, T., Khedher, L., Toumi, H., & Lespessa-illes, E. (2019). A decision support tool for early detection of knee OsteoArthritis using X-ray imaging and machine learning: Data from the OsteoArthritis Initiative. *Computerized Medical Imaging and Graphics*, *73*, 11–18. DOI: 10.1016/j.comp-medimag.2019.01.007 PMID: 30784984

Cueva, J. H. (2022). Detection and Classification of Knee Osteoarthritis. *MDPI*, 2362.

Cui, A., Li, H., Wang, D., Zhong, J., Chen, Y., & Lu, H. (2020). Global, regional prevalence, incidence and risk factors of knee osteoarthritis in population-based studies. *EClinicalMedicine*, *29-30*, 29–30. DOI: 10.1016/j.eclinm.2020.100587 PMID: 34505846

Du, Y. (2018). A Novel Method to Predict Knee Osteoarthritis Progression on MRI Using Machine Learning Methods. *IEEE*, 228-236.

Du, Y. (2018). Knee osteoarthritis severity level classification using whole knee cartilage damage Index and ANN. *CHASE '18: Proceedings, International Conference on Connected Health: Applications, Systems and Engineering Technologies.* IEEE/ACM. DOI: 10.1145/3278576.3278585

Du, Y., Almajalid, R., Shan, J., & Zhang, M. (2018). A Novel Method to Predict Knee Osteoarthritis Progression on MRI Using Machine Learning Methods. *IEEE Transactions on Nanobioscience*, *17*(3), 228–236. DOI: 10.1109/TNB.2018.2840082 PMID: 29994316

El-Ghany, S. A. (2023). A fully automatic fine tuned deep learning model for knee osteoarthritis detection and progression analysis. *Egyptian Informatics Journal*, 229-240.

*IR(Infrared) Knee Joint Dataset*. (n.d.). Retrieved 5 28, 2024, from Mrinal Kanti Bhowmik: https://mkbhowmik.in/irD.aspx

Jamshidi, A., Leclercq, M., Labbe, A., Pelletier, J.-P., Abram, F., Droit, A., & Martel-Pelletier, J. (2020). Identification of the most important features of knee osteoarthritis structural progressors using machine learning methods. *Therapeutic Advances in Musculoskeletal Disease*, *12*, 1759720X20933468. DOI: 10.1177/1759720X20933468 PMID: 32849918

*Kaggle*. (n.d.). Retrieved 5 28, 2024, from Knee Osteoarthritis Dataset with Severity Grading: https://www.kaggle.com/datasets/shashwatwork/knee-osteoarthritis -dataset-with-severity

Kanthavel, R. (2022). *Prediction Model Using Reinforcement Deep Learning Technique for*. Tech Science Press.

Kwon, S. B. (2020). *Machine Learning-Based Automatic Classification of Knee Osteoarthritis Severity Using Gait Data and Radiographic Images*. IEEE. DOI: 10.1109/ACCESS.2020.3006335

Mahum, R. (2021). A Novel Hybrid Approach Based on Deep CNN Features to Detect Knee Osteoarthritis. *MDPI*, 6189.

Moustakidis, S., Christodoulou, E., Papageorgiou, E., Kokkotis, C., Papandrianos, N., & Tsaopoulos, D. (2019). Application of machine intelligence for osteoarthritis classification: A classical implementation and a quantum perspective. *Quantum Machine Intelligence*, *1*(3-4), 73–86. DOI: 10.1007/s42484-019-00008-3

*National Institute of Health*. (n.d.). Retrieved 12 31, 2023, from The Osteoarthritis Initiative: https://nda.nih.gov/oai

Olsson, S. (2021). Automating classifcation of osteoarthritis. *BMC Musculoskeletal Disorders*.

Ondrésik, M. (2016). Management of knee osteoarthritis. Current status and future trends. *Biotechnology and Bioengineering*, 717–739. PMID: 27618194

Piccolo, C. L., Mallio, C. A., Vaccarino, F., Grasso, R. F., & Zobel, B. B. (2023). Imaging of knee osteoarthritis: A review of multimodal diagnostic approach. *Quantitative Imaging in Medicine and Surgery*, *13*(11), 7582–7595. DOI: 10.21037/qims-22-1392 PMID: 37969633

Pollard, T. C. (2008). The assessment of early osteoarthritis. *The Bone & Joint*, 411-421.

Pongsakonpruttikul, N., Angthong, C., Kittichai, V., Chuwongin, S., Puengpipattrakul, P., Thongpat, P., & Tongloy, T. (2022). Artificial intelligence assistance in radiographic detection and classification of knee osteoarthritis and its severity: A cross-sectional diagnostic study. *European Review for Medical and Pharmacological Sciences*, *26*(5), 1549–1558.

Rahmati, M., Mobasheri, A., & Mozafari, M. (2016). Inflammatory mediators in osteoarthritis: A critical review of the state-of-the-art, current prospects, and future challenges. *Bone*, *85*, 81–90. DOI: 10.1016/j.bone.2016.01.019 PMID: 26812612

*Sethu Institute of Technology*. (n.d.). Retrieved 5 28, 2024, from OSTHERSIT: https://sethu.ac.in/osthersit/

Spitaels, D. (2020). Epidemiology of knee osteoarthritis in general practice: A registry-based study. *BMJ (Clinical Research Ed.)*. PMID: 31964664

Tariq, T. (2023). Knee Osteoarthritis Detection and Classification Using X-Rays. *IEEE*, 48292 - 48303.

Wang, Y. (2021). *An Automatic Knee Osteoarthritis Diagnosis Method Based on Deep Learning: Data from the Osteoarthritis Initiative*. Hindawi.

Zebari, D. A. (2022). Knee Osteoarthritis Detection Using Deep Feature Based on Convolutional Neural Network. *International Conference on Computer Science and Software Engineering (CSASE)* (pp. 259-264). IEEE. DOI: 10.1109/CSASE51777.2022.9759799

# Chapter 2
# Automatic Hybrid CNN-Based Skin Cancer Classification


**Singaravelan Shanmugasundaram**
https://orcid.org/0000-0003-4353-2261
*PSR Engineering College, India*

**Arun Shunmugam D.**
https://orcid.org/0000-0002-1325-9017
*PSR Engineering College, India*

**Anjel Jean Vincy K.**
*Vel Tech Rengarajan Dr. Sagunthala R&D Institute of Science and Technology, India*

**Vijaya Rani G.**
*Francis Xavier Engineering College, India*

**I. Noormohamed**
*R.M.K. Engineering College, India*

**Suresh Chinnathampy M.**
*Vel Tech High Tech Dr. Rangarajan Dr. Sakunthala Engineering College, India*

**M. Dhivya**
*Ramco Institute of Technology, India*

**G. Mareeswari**
*Ramco Institute of Technology, India*



## ABSTRACT

*Skin cancer, particularly dermo-cancer, is a critical health concern with rising incidences worldwide. Automated classification of dermo-cancer from skin images plays a pivotal role in early diagnosis and timely intervention. In this work, hybrid architecture that integrates inception and ResNet models to enhance feature extraction and facilitate hierarchical learning for improved dermo-cancer classification is explored. The inception module contributes to capturing multi-scale features, while the ResNet module addresses the challenges of vanishing gradients and aids in building a more robust and deeper neural network. The proposed hybrid architec-*



DOI: 10.4018/979-8-3693-3840-7.ch002

*ture is trained on a comprehensive dataset, and experimental results demonstrate superior performance compared to individual models, achieving enhanced accuracy, sensitivity, and specificity. The approach automated dermo-cancer classification but also holds promise for other medical image tasks, showcasing the potential of hybrid architectures in medical image analysis.*

## INTRODUCTION

Cancer is when cells in the body change and grow out of control. To help you understand what happens when you have cancer, let's look at how your body works normally. Your body is made up of tiny building blocks called cells. Normal cells grow when your body needs them, and die when your body does not need them any longer.

Cancer is made up of abnormal cells that grow even though your body doesn't need them. In most cancers, the abnormal cells grow to form a lump or mass called a tumor. If cancer cells are in the body long enough, they can grow into (invade) nearby areas. They can even spread to other parts of the body (metastasis).

Skin cancer is a disease that begins in the cells of the skin. The area of skin with the cancer is often called a lesion. There are several types of skin cancer (carcinoma). Melanoma is the most serious. But there are others that are known as nonmelanoma skin cancer. These include:

- Basal cell carcinoma
- Squamous cell carcinoma
- Merkel cell carcinoma
- Cutaneous T-cell lymphoma
- Kaposi sarcoma

Basal cell carcinoma and squamous cell carcinoma are by far the most common.

### 1.1. Understanding the Skin

The skin is the largest organ of the body. Skin protects us from heat, sunlight, injury, and infection. It also stores water and fat, and makes vitamin D. The skin has 3 layers:

- The outer layer called the epidermis
- The middle layer called the dermis
- The inner layer called the subcutis (subcutaneous)

The epidermis is made of flat cells called squamous cells. Round basal cells are under the squamous cells. The lower part of the epidermis has pigment-producing cells called melanocytes. These cells darken the skin when exposed to the sun (Kang & Wang, 2014).

The dermis has blood vessels, lymphatic vessels, hair follicles, and glands. Some of these glands make sweat, which helps keep the body cool. Other glands make an oily substance called sebum. Sebum helps keep the skin from getting dry. Sweat and sebum reach the skin's surface through tiny openings called pores (Long et al., 2015). The subcutis and the lowest part of the dermis form a network of collagen and fat cells. This layer conserves heat and helps protect the body's organs from injury.

## 1.2. Different Types of Nonmelanoma Skin Cancer

### Basal Cell Carcinoma

Basal cell carcinoma, also known as basal cell cancer, is the most common type of skin cancer. It begins in basal cells in the deepest part of the epidermis. It often starts in areas of skin exposed to the sun, such as the face, head, neck, arms, and hands. The cancer lesion often appears as small, raised, shiny, or pearly bumps, but it can have various kinds of appearance. They tend to grow slowly and rarely spread to other parts of the body

### Squamous Cell Carcinoma

Squamous cell carcinoma, also known as squamous cell cancer, is the second most common type of skin cancer. It begins in flat cells called squamous cells in the upper part of the epidermis. Like basal cell cancer, it often starts in areas of skin exposed to the sun, such as the face, head, neck, arms, and hands. But it can also start in other parts of the body, such as skin in the genital area.

### Merkel Cell Carcinoma

Merkel cell cancer is a rare type of skin cancer. Merkel cells are types of cells in the upper layer of the skin. The cells are very close to nerve endings, and help the skin sense light touch. Merkel cell cancer occurs when these cells grow out of control. Merkel cell cancer can be dangerous because it tends to grow quickly (Ronneberger et al., 2015). It can be hard to treat if it spreads beyond the skin. Merkel cell cancer tumors are most often found on sun-exposed areas of skin, such as the face, neck, and arms. But they can start anywhere on the body. They usually appear as firm, shiny skin lumps that do not hurt. The lumps may be red, pink, or blue. They tend to grow very quickly.

### Cutaneous T-Cell Lymphoma

Cutaneous T-cell lymphoma is a type of cancer that starts in blood cells called T- lymphocytes. These are white blood cells that are part of your immune system. They normally fight infection in the body. The cancer then affects the skin (cutaneous). It causes scaly patches or bumps. The cancer is also known as lymphoma of the skin. It is a type of non-Hodgkin lymphoma. It is usually a slow-growing cancer (Huang & Jain, 2013). It develops over many years. The two most common types of this cancer are mycosis fungoides and the Sezary syndrome.

## 1.3. Melanoma Skin Cancer

Melanoma is a form of skin cancer that occurs in the melanocytes, which are cells in the outer layer of the skin (the epidermis). Melanocytes produce the skin coloring or pigment known as melanin, which gives skin its tan or brown colour and helps protect the deeper layers of the skin from the harmful effects of the sun (Amaral et al., 2014). Melanoma develops when melanocytes undergo malignant transformation, become abnormal, grow uncontrollably and aggressively invade surrounding tissues. Melanoma may affect only the skin, or it may spread through the blood or lymph system to other organs and bones.

Melanoma is the most serious form of skin cancer. Melanoma may be cured if caught and treated early, but, if left untreated, the majority of melanomas eventually spread to other parts of the body. Early detection and surgery to remove the melanoma are successful in curing most cases of melanoma; however, it is rarely curable in its later stages.

In the United States, melanoma accounts for approximately 4% of all newly diagnosed cancers annually. The American Cancer Society estimates that in 2004, approximately 55,100 new melanomas will be diagnosed and about 7,910 deaths due to melanoma will occur in the United States (Badrinarayanan et al., 2015).

Internationally, the incidence of melanoma varies greatly, with the highest incidence occurring in Australia, the United States, Norway, Switzerland, Sweden, Denmark and Israel, and the lowest incidence in Japan, the Philippines, China and India.

Melanoma commonly afflicts the young and middle aged; however, people of all ages are at risk. It is the most common cancer in young adults aged 20-30 and is the leading cause of cancer death for women aged 25-30. Melanoma is significantly more prevalent among white populations than in blacks and Asians; the incidence of melanoma in blacks is approximately 1/20 than that of whites.

There has been a dramatic increase in melanoma incidence over the last century. The lifetime risk of developing melanoma in 1935 was only 1 per 1500 Americans, and the lifetime risk for melanoma in 2004 was 1 in 71 Americans (Long et al., 2015). The lifetime risk is estimated to rise to 1 in 50 Americans by 2010. Fortunately, due to prevention and early detection practices, melanoma mortality rates have not increased as sharply and have remained stable or decreased since the 1990s.

## 1.4. Types of Melanomas

The appearance and growth of melanoma will differ depending on the morphologic type:

### *Superficial spreading melanoma*

Superficial spreading melanoma is the most common type of melanoma, representing approximately 70% of all melanomas. This melanoma goes through a prolonged (years) horizontal growth pattern on the skin before becoming invasive. Superficial spreading melanomas are flat or slightly elevated brown lesions with black, blue or pink discolouration which are typically greater than 6 mm in diameter and have irregular asymmetric borders. These melanomas may be found on any body surface, especially the head, neck, and trunk of males and the lower extremities of females (Houssein et al., 2024).

### *Nodular melanoma*

Nodular melanoma represents about 15% of all melanomas and becomes invasive soon after first appearing. A nodular melanoma typically looks similar to a blood vessel growth, presenting as a dark brown-to-black papule or dome-shaped nodule; however, 5% of nodular melanomas are amelanotic (see below). These melanomas are commonly found on all body surfaces, especially the trunk of males.

### *Acral-lentiginous melanoma*

Acral-lentiginous melanoma represents approximately 8% of all melanomas and is the most common melanoma in dark-skinned people. Acral-lentiginous melanomas represent up to 70% of melanomas in blacks and up to 46% in Asians. This type can occur on the palms, soles and nail beds (subungual). Like nodular melanoma, acral-lentiginous melanoma is extremely aggressive, with rapid progression from the horizontal to vertical growth phase.

#### *Lentigo maligna melanoma*

Lentigo maligna melanoma accounts for approximately 5% of melanomas. Lentigo maligna melanomas are typically found on sun-exposed areas of the skin in adults and are clearly linked to exposure to the sun. Often many years pass between the first appearance of this melanoma and when it becomes invasive. The precursor lesion is typically greater than 3 cm in diameter, and, upon becoming invasive, develops a dark brown-to-black colour or raised blue-black nodule.

#### *Amelanotic melanoma*

Amelanotic Melanoma is rare and challenging to diagnose because there is an absence of pigmentation (colour). However, hallmark traits of melanoma, such as changes in size, borders, and symmetry, are present in this melanoma.

#### *Desmoplastic melanoma*

Desmoplastic Melanoma is rare, representing approximately 1.7% of all melanomas. This type of melanoma is locally aggressive and difficult to diagnose both clinically and microscopically. The majority of these tumors occur on the head and neck of elderly patients and one half are amelanotic.

## 1.5. Risk Factors

The usual risk factors associated with cancer, e.g., cigarette smoking, are not associated with melanoma risk. Rather, melanoma risk factors include sun exposure, family history, genetics and phenotypical traits such as skin, hair and eye colour.

Known risk factors for melanoma of the skin include:

- Previous melanoma or other skin cancer
- Family history of melanoma
- Atypical moles (dysplastic nevi)
- 50 or more moles
- History of exposure to ultraviolet (UV) radiation, including UV exposure at tanning salons
- History of sunburns, particularly severe sunburns in childhood
- White race, especially people who have fair skin that burns rather than tans

- Red or blonde hair
- Blue eyes
- Many (extensive) freckles on the upper back
- Other cancers such as leukemia or lymphoma
- Impaired immune system
- Moles that are present at birth, especially if the moles are larger than 20 cm (giant congenital melanocytic nevus)
- Equatorial latitudes
- PUVA treatment, used to treat skin conditions such as psoriasis

## 1.6. Signs and Symptoms

Change in an existing mole is the most common symptom of melanoma. Education regarding the warning signs of early melanoma has been achieved through the use of the American Cancer Society's ABCD system of detection, which includes looking for changes in the following attributes of a mole:

***Asymmetry***

One half of the mole or skin growth doesn't match the other half.

***Border irregularity***

The edges of the mole or skin growth are ragged, notched, or blurred.

***Colour***

The pigmentation of the growth is not uniform. Shades of tan, brown, and black are present. Dashes of red, white, and blue may create a mottled appearance. A part of the mole may lose its colour. Changes in colour distribution, especially the spread of colour from the edge of a mole into the surrounding skin may occur.

***Diameter***

The mole or skin growth is larger than 6 mm (0.2 in.), or about the size of a pencil eraser.

Size may increase rapidly; however, any growth of a mole should be of concern. Other signs of melanoma in a mole include changes in:

- **Elevation:**

A previously flat mole may thicken or rise up.

- **Surface:**

The mole may become scaly, erode, ooze, bleed, become crusted or ulcerate.

- **Surrounding**

Skin around the mole may become red or swollen, or develop small new patches of colour around a larger lesion (satellite pigmentations).

- **Sensation:**

The mole may begin to itch, burn, or become painful.

- **Consistency:**

The mole may soften and small pieces of skin may break off easily (friability).

Photographs are sometimes used to document and detect changes in the skin, especially a typical mole. Computers may be utilized to compare photographs of suspicious moles and lesions taken from exam to exam to more accurately determine whether a mole or lesion is changing.

## 1.7. Skin Cancer Detection

Melanoma is considered the deadliest form of skin cancer and is caused by the development of a malignant tumor of the melanocytes. The objective of the skin cancer detection project is to develop a framework to analyze and assess the risk of melanoma using dermatological photographs taken with a standard consumer-grade camera.

The skin cancer detection framework consists of novel algorithms to perform the following:

- illumination correction preprocessing
- segmentation of the lesion
- feature extraction

Our data set is provided at the end of the page. This includes images extracted from the public databases DermIS and DermQuest, along with manual segmentations of the lesions.

## Correction Preprocessing

The first step in the proposed framework is a preprocessing step, where the image is corrected for illumination variation (i.e. the presence of shadows and brightly illuminated areas). Our approach to the problem of correcting is a multi-stage illumination modeling algorithm. We assume an illumination-reflectance multiplicative model where each pixel from the Value colour channel (in the HSV colour space) can be decomposed into an illumination component and a reflectance component. The goal of the algorithm is to first estimate the illumination component and calculate the reflectance component using that illumination estimate (Singaravelan et al., 2024).

In the VIP lab, the proposed multi-stage illumination modelling algorithm uses the following stages to estimate and correct for illumination variation (shown in Figure 1):

- Initial Monte Carlo illumination estimate
- Final parametric illumination estimate
- Calculate the reflectance component

*Figure 1. Flow chart showing the steps to correct the input image for illumination variation.*

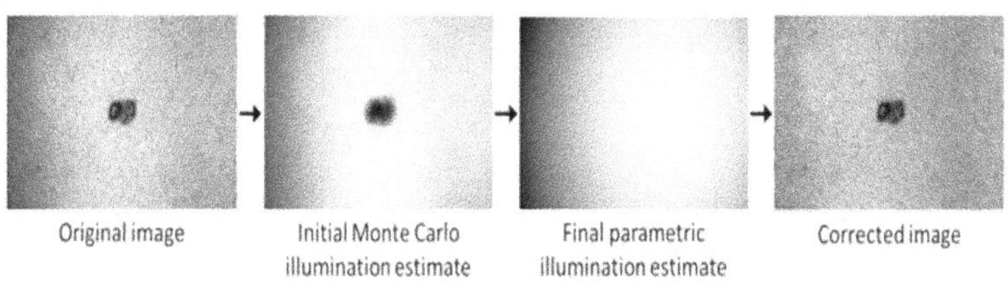


## Skin Lesion Segmentation

The objective of the skin lesion segmentation step is to find the border of the skin lesion. It is important that this step is performed accurately because many features used to assess the risk of melanoma are derived based on the lesion border. Our approach to finding the lesion border is a texture distinctiveness-based lesion segmentation (Prasad et al., 2023).

*Figure 2. Flow chart showing the steps to calculate the texture distinctiveness of metric*

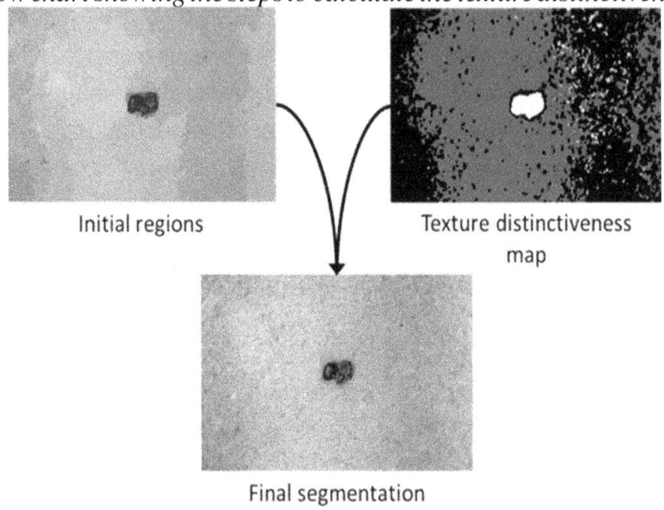


The first stage of the skin lesion segmentation algorithm is to learn representative texture distributions and calculate the texture distinctiveness metric for each distribution (shown in Figure 2; Behara et al., 2024). A texture vector is extracted for each pixel in the image. Then, a Gaussian mixture model is implemented that learns the texture distributions based on the set of texture vectors. Finally, the dissimilarity between a texture distribution and all other texture distirbutions is measured, which is quantified as the texture distinctiveness metric.

*Figure 3. Flow chart showing the merging of the texture distinctiveness map and initial regions. Each solid colour in the image of initial regions corresponds to a single region.*

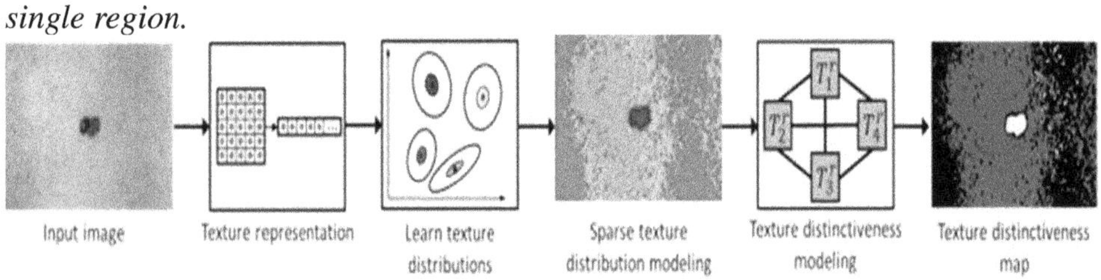


In the second stage, the pixels in the image are classified as being part of the normal skin or lesion class (shown in Figure 3). To do this, the image is divided into a number of regions. These regions are combined with the texture distinctiveness map to find the skin lesion (Lee, 1994).

### Feature Extraction

In order to use classification techniques, the image must be transformed such that it represented a point in some *n*-dimensional feature space. The axes in this feature space represent calculations that are relevant to describing the observed phenomenon (e.g., malignancy).

## 2. LITERATURE REVIEW

### Ultrawideband, Stable Normal and Cancer Skin Tissue Phantoms for Millimeter-Wave Skin Cancer Imaging

Amir Mirbeik et al. (2018) introduced new, stable, and broadband skin-equivalent semisolid phantoms for mimicking interactions of millimeter waves with the human skin and skin tumors. Realistic skin phantoms serve as an invaluable tool for exploring the feasibility of new technologies and improving design concepts related to millimeter-wave skin cancer detection methods. Normal and malignant skin tissues are separately mimicked by using appropriate mixtures of deionized water, oil, gelatine powder, formaldehyde, TX-150 (a gelling agent, widely referred to as "super stuff"), and detergent. The dielectric properties of the phantoms are characterized over the frequency band of 0.5-50 GHz using a slim-form open-ended coaxial probe in conjunction with a millimeter-wave vector network analyser. The measured permittivity results show excellent match with ex vivo, fresh skin (both normal and malignant) permittivity's determined in our prior work over the entire frequency range.

### Automated Detection and Segmentation of Vascular Structures of Skin Lesions Seen in Dermoscopy, With an Application to Basal Cell Carcinoma Classification

Pegah Kharazmi et al. (2017) proposed a novel framework for detection and segmentation of cutaneous vasculature from dermoscopy images and the further extracted vascular features are explored for skin cancer classification. Given a dermoscopy image, we segment vascular structures of the lesion by first decomposing the image using independent component analysis into melanin and hemoglobin components. This eliminates the effect of pigmentation on the visibility of blood vessels. Using k-means clustering, the hemoglobin component is then clustered into normal, pigmented, and erythema regions. Shape filters are then applied to the erythema cluster at different scales.

## Water-Based Terahertz Metamaterial for Skin Cancer Detection Application

Afsaneh Keshavarz et al. (2018) presented a highly sensitive detection of skin cancer using a water-based terahertz (THz) metamaterial (MM) and a semiconductor film. This paper applied the application of terahertz pulsed imaging in reflection geometry for the study of skin tissue and related cancers. As a refractive index (RI) sensing application of the proposed device, with placing different sensing materials in the biosensor design, the effective RI will also change that in turn leads to measuring the sensitivity of the biosensor to detect the normal skin and basal cell carcinoma. The RI figure of merit (FOM) value of the proposed device is much higher than that of the sensor using a semiconductor film to detect biomarkers in the literature.

## Noninvasive Real-Time Automated Skin Lesion Analysis System for Melanoma Early Detection and Prevention

Omar Abuzaghleh et al. (2015) proposed the two major components of a non-invasive real-time automated skin lesion analysis system for the early detection and prevention of melanoma. The first component is a real-time alert to help users prevent skinburn caused by sunlight; a novel equation to compute the time for skin to burn is thereby introduced. The second component is an automated image analysis module, which contains image acquisition, hair detection and exclusion, lesion segmentation, feature extraction, and classification. The proposed system uses PH2 Dermoscopy image database from Pedro Hispano Hospital for the development and testing purposes. The image database contains a total of 200 dermoscopy images of lesions, including benign, atypical, and melanoma cases.

## Synthetic Ultra-High-Resolution Millimeter-Wave Imaging for Skin Cancer Detection

Amir Mirbeik et al. (2018) introduced, for the first time, a millimeter-wave imaging system with a "synthetic" ultrawide imaging bandwidth of 98 GHz to provide the ultra-high resolutions required for early-stage skin cancer detection. The proposed approach consists of splitting the required ultra-wide imaging bandwidth into four sub-bands, and assigning each sub-band to a separate imaging element, i.e., an antenna radiator. Each of the sub-band antennas transmits and receives signals only at its corresponding sub-band. The captured signals are then combined and processed to form the image of the target. For each sub-band, a Vivaldi tapered slot antenna fed with a combination of substrate-integrated waveguide and coplanar waveguide is designed and microfabricated. Design techniques are also provided for the four similarly- shaped sub-band antennas for achieving excellent impedance matches ($S_{11}$ <; -10 dB) and nearly constant gains of 10 dBi over the entire 12- 110 GHz bandwidth. The design procedure is validated by comparing the simulated results with measurements performed on the fabricated prototypes.

## A New Total Body Scanning System for Automatic Change Detection in Multiple Pigmented Skin Lesions

Konstantin Korotkov et al. (2015) proposed a photogrammetry-based total body scanning system allowing for skin surface image acquisition using cross-polarized light. Equipped with 21 high-resolution cameras and a turntable, this scanner automatically acquires a set of overlapping images, covering 85%-90% of the patient's skin surface. These images are used for the automated mapping of PSLs and their change estimation between explorations. The maps produced relate images of individual lesions with their locations on the patient's body, solving the body-to-image and image-to-image correspondence problem in TBSEs. Currently, the scanner is limited to patients with sparse body hair and, for a complete skin examination, the scalp, palms, soles and inner arms should be photographed manually.

## Accessible Melanoma Detection Using Smartphones and Mobile Image Analysis

Thanh-Toan Do et al. (2015) investigated the design of an entire mobile imaging system for early detection of melanoma. Different from previous work, also focus on smartphone- captured visible light images. Our design addresses two major challenges. First, images acquired using a smartphone under loosely-controlled environmental conditions may be subject to various distortions, and this makes melanoma detection more difficult. Second, processing performed on a smartphone is subject to stringent computation and memory constraints. This work proposed a detection system that is optimized to run entirely on the resource-constrained smartphone. This system intends to localize the skin lesion by combining a lightweight method for skin detection with a hierarchical segmentation approach using two fast segmentation methods. Moreover, this was an extensive set of image features and propose new numerical features to characterize a skin lesion. Furthermore, we propose an improved feature selection algorithm to determine a small set of discriminative features used by the final lightweight system. In addition, also study the human-computer interface (HCI) design to understand the usability and acceptance issues of the proposed system.

## In Vitro Dielectric Properties of Rat Skin Tissue for Microwave Skin Cancer Detection

Cemanur Aydinalp et al. (2019) developed the current measurement techniques and tools has to be optimized for skin cancer detection. To this end, this study presents dielectric property measurements with open-ended coaxial probes having small apertures customized for detection of skin cancer.

## Detection Skin Cancer Using SVM and Snake Model

Prachya Bumrungkun et al. (2018) extracted and analyzed such features, image segmentation plays important role for automatic skin cancer detection system. In this paper, we propose the image segmentation scheme based on Support Vector Machine (SVM) and Snake active contour. SVM is used to help finding the appropriate parameters for snake algorithm.

# 3. SYSTEM ARCHITECTURE

## 3.1 Existing System

Automatic image segmentation techniques can be classified into four categories, namely, (1) Clustering Methods, (2) Thresholding Methods, (3) Edge-Detection Methods, and

(4) Region-Based Methods [6].

**Clustering Methods**

Clustering is a process whereby a data set (pixels) is replaced by cluster; pixels may belong together because of the same colour, texture etc. There are two natural algorithms for clustering: divisive clustering and agglomerative clustering. The difficulty in using either of the methods directly is that there are lots of pixels in an image. Also, the methods are not explicit about the objective function that is being optimized. An alternative approach is to write down an objective function and then build an algorithm (Cancer Council Australia, 2010). The K-means algorithm is an iterative technique that is used to partition an image into K clusters, where each pixel in the image is assigned to the cluster that minimizes the variance between the pixel and the cluster center and is based on pixel colour, intensity, texture, and location, or a weighted combination of these factors. This algorithm is guaranteed to converge, but it may not return the optimal solution. The quality of the solution depends on the initial set of clusters and the value of K.

### Thresholding Methods

Thresholding is the operation of converting a multilevel image into a binary image i.e., it assigns the value of 0 (background) or 1 (objects or foreground) to each pixel of an image based on a comparison with some threshold value T (intensity or colour value). When T is constant, the approach is called global thresholding; otherwise, it is called local thresholding. Global thresholding methods can fail when the background illumination is uneven. Multiple thresholds are used to compensate for uneven illumination. Threshold selection is typically done interactively.

### Edge Detection Methods

Edge detection methods locate the pixels in the image that correspond to the edges of the objects seen in the image. The result is a binary image with the detected edge pixels. Common algorithms used are Sobel, Prewitt, Robert, Canny and Laplacian operators. These algorithms are suitable for images that are simple and noise free; and will often produce missing edges, or extra edges on complex and noisy images.

### Region-Based Methods

The goal of region-based segmentation is to use image characteristics to map individual pixels in an input image to sets of pixels called regions that might correspond to an object or a meaningful part of one. The various techniques are: Local techniques, Global techniques and splitting and merging techniques. The effectiveness of region growing algorithms depends on the application area and the input image. If the image is sufficiently simple, simple local techniques can be effective. However, on difficult scenes, even the most sophisticated techniques may not produce a satisfactory segmentation (Green & Martin, 1994). Edge-based techniques are based on the assumption that pixel values change rapidly at the edge between two regions. Operators such as Sobel or Roberts operators can be used to detect the edges. And some post procedures such as edge tracking, gap filling can be used to generate closed curves. Region based techniques are based on the assumption that adjacent pixels in the same region should be consistent in some properties. Namely, they may have similar characteristic such as grey value, colour value or texture. The deformable models are based on curves or surfaces defined within an image that moves due to the influence of certain forces. And the global optimization approaches use a global criterion when segmenting the image.

## 3.2 Proposed System

In this work, we propose a novel Hybrid Architecture integrating Inception and ResNet models to enhance feature extraction and improve dermo-cancer classification.

*Figure 4. Proposed system*

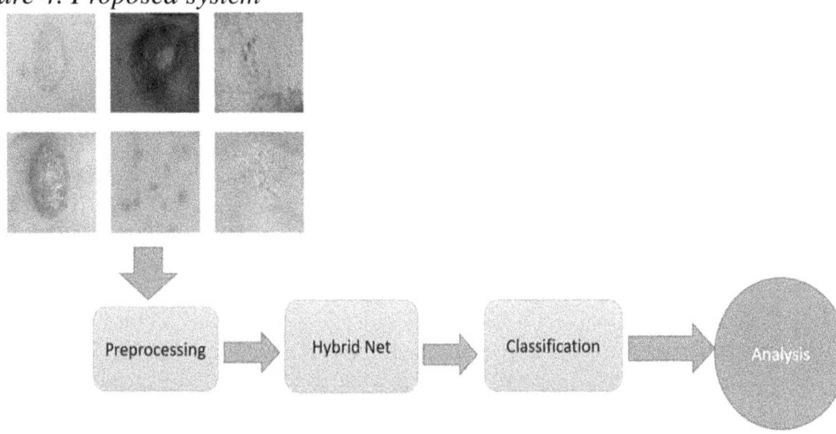


Our Hybrid Architecture combines the strengths of the Inception and ResNet models. The Inception module captures multi-scale features, enabling intricate pattern recognition in skin lesions. Concurrently, the ResNet module addresses challenges such as vanishing gradients, contributing to the development of a deeper and more robust neural network. By fusing these architectures, our system achieves synergistic effects, leveraging the complementary features of both models.

*Figure 5. Hybrid architecture*

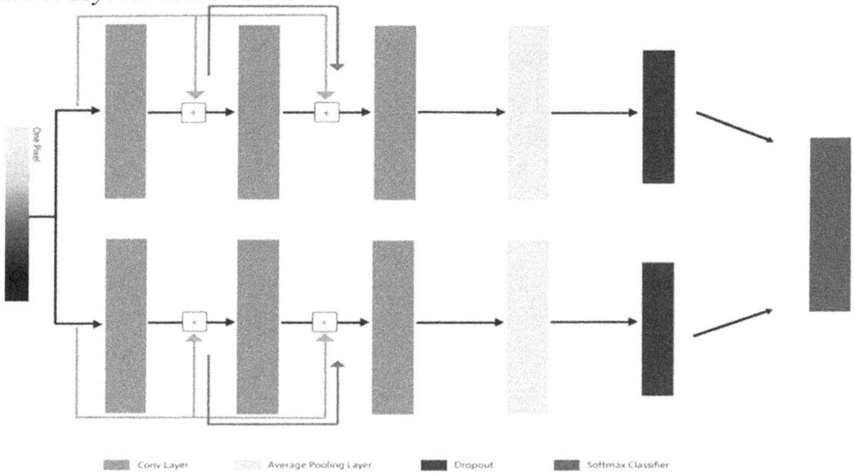


Key Modules:

***Inception Module:***

The Inception module, a cornerstone of our proposed Hybrid Architecture, is renowned for its ability to capture multi-scale features efficiently. This module consists of a series of convolutional layers with varying kernel sizes, enabling it to analyze input images at different scales simultaneously. The architecture of the Inception module comprises several parallel convolutional pathways, each processing the input data using different filter sizes. These pathways are then concatenated to form the module's output, allowing for the extraction of features at multiple scales within a single layer.

Within each pathway of the Inception module, convolutional layers with different filter sizes are employed to capture features of varying spatial extents. This design choice enables the module to discern both fine-grained details and broader patterns within the input images. Additionally, the inclusion of pooling layers helps in down sampling the feature maps, reducing computational complexity while retaining essential information.

Furthermore, to enhance the representational capacity of the module, 1x1 convolutions are incorporated to perform dimensionality reduction. This step reduces the number of input channels before applying larger filter sizes, thereby improving computational efficiency without sacrificing expressive power. Batch normalization and activation functions such as ReLU are also employed to stabilize training and introduce non-linearity into the model.

The output of each pathway within the Inception module is concatenated along the channel dimension, resulting in a feature map that encapsulates a diverse range of spatial information. This aggregated representation is then passed to subsequent layers for further processing or classification. By leveraging the strengths of parallel convolutional pathways with varying filter sizes, the Inception module excels at capturing intricate patterns and structures within skin lesions, thereby facilitating accurate dermo-cancer classification.

In summary, the Inception module serves as a pivotal component of our Hybrid Architecture, providing the capability to extract multi-scale features efficiently. Through its sophisticated design incorporating parallel convolutional pathways and dimensionality reduction techniques, the module enables our system to discern subtle abnormalities in skin images, contributing to the accurate classification of dermo-cancer.

***ResNet Module:***

The ResNet module, an integral part of our proposed Hybrid Architecture, is renowned for its ability to address the challenges of training very deep neural networks. Its unique architecture, featuring residual connections, enables the effective propagation of gradients and facilitates the training of deeper models without suffering from the vanishing gradient problem.

At the heart of the ResNet module lies the concept of residual learning, where the input to a layer is combined with its output using a shortcut connection. This skip connection bypasses one or more layers, allowing the network to learn residual functions, i.e., the difference between the input and output of a layer. This design choice enables the network to focus on learning the residual information, which is often easier than directly learning the desired mapping.

The ResNet architecture typically consists of several residual blocks, each containing multiple convolutional layers. Within each residual block, the input undergoes a series of convolutions, batch normalization, and activation functions, similar to traditional convolutional neural networks. However, the key innovation lies in the addition of the shortcut connection, which directly adds the input to the output of the block before applying the activation function. This additive operation preserves the gradient flow and alleviates the vanishing gradient problem, enabling more effective training of deeper networks (Aitken et al., 1996).

Moreover, to further enhance the representational capacity of the network, bottleneck blocks are often employed within ResNet architectures. These bottleneck blocks consist of a sequence of 1x1, 3x3, and 1x1 convolutions, where the 1x1 convolutions are used for dimensionality reduction and expansion. This design reduces the computational cost while maintaining the expressive power of the network, allowing for the efficient training of very deep models.

Additionally, batch normalization and activation functions such as ReLU are employed within each residual block to stabilize training and introduce non-linearity into the model. These components play a crucial role in accelerating convergence and improving the overall performance of the network.

In summary, the ResNet module revolutionizes deep neural network architectures by introducing residual connections, which facilitate the training of extremely deep models (Chang et al., 2005). Through its innovative design and use of residual learning, the module enables our Hybrid Architecture to build robust and deeper neural networks, thus enhancing the classification performance for dermo-cancer detection.

***Experimental Validation:***

The proposed Hybrid Architecture is trained on a comprehensive dataset comprising diverse skin lesions. Experimental results demonstrate superior performance compared to individual models, with enhanced accuracy, sensitivity, and specificity in dermo- cancer classification tasks.

***Significance:***

Our approach not only advances automated dermo-cancer classification but also holds promise for other medical image analysis tasks. The utilization of hybrid architectures showcases the potential for integrating diverse models to improve the accuracy and effectiveness of medical image classification systems.

***Architecture:***

In contrast to the existing deep learning methods, we pro pose a new CNN model that takes advantage of both the Inception model and the deep residual network ResNet model by adopting and combining the core ideas of both ResNet and Inception into one model. Deep residual networks and the Inception model have proven their ability to scale up thousands of layers while providing improved efficiency and better performance. The residual networks consist of many residual blocks with identity mapping. The Inception model is a deep convolutional network that con sists of many convolutional networks. One objective of HSI classification is to predict the land cover type by assigning and labeling individual pixels, which have many frequency bands, into individual classes. In this paper, we present a deep hyper-network architecture that can learn the deep features of HSI and provide good performance without any type of dataset augmentation or many preprocessing steps. Figure 3 shows our final proposed hybrid deep ResNet-Inception network architecture. The proposed architecture consists of two residual blocks, as shown in Figure

The network has three convolutional layers followed by one average pooling layer. The outputs of each layer form the inputs to each successive layer. The architecture consists of one fully connected cascaded residual block, in this residual model, each convolutional layer receives inputs from all the previous convolutional layers. We determined the number of convolutional layers empirically and found that three convolutional layers are optimal for our model. The convolutional layers apply convolution operations to the input data, and the last pooling layer applies an average pooling operation to the data before feeding it to the classifier (She et al., 2007). We use the Adam optimization algorithm, instead of the classical stochastic gradient descent to optimize our network because the former provides faster convergence. Furthermore, the Adam optimization algorithm is computationally efficient and less sensitive to noise. The initial learning rate is 0.001, and the batch size is set to 17 for all our datasets (the University of Pavia dataset, the Salinas dataset and the Pavia Centre scene dataset).

### *Convolutional layers*

The convolutional layers apply a convolution operation to the data and then transform the data using the rectified linear unit (ReLU) function. There are three convolutional layers, each of which has 9 filters, that output 9 feature maps. The operation of each kernel (filter) is defined in equation:

$$Xi = (Wi * Xi-1 + i)$$

where $*$ is the convolution operator, that convolves the filter Wi with the input data Xi−1, adds the bias term, and then applies the rectifier function and outputs a feature map Xi. In our proposed model, the size of each convolutional filter (1D convolution window) is 16 units, and there are 9 filters in each layer. Each filter pads the input such that the output has the same dimensions as the input tensor. The convolution stride length is one. The weights (kernels) of the convolutional layers are initialized using the Glorot uniform (Xavier) weight initialization method as suggested in (Glorot and Bengio 2010). The bias terms are initialized to zeros. The ReLU activation function applies an elementwise operation on the input data x as defined in equation:

$$(x)=\max(x,0) \qquad (2)$$

We used 1D convolutional kernels instead of 2D or 3D ker nels because the former are more suitable for the structure of the HSI data, in which each pixel and its band are repre sented as one vector that has one label. Our model includes two residual models that are ulti mately connected. The operation of the upper residual model is defined in equation:

$$X1 = (W1 * X0 + 1)$$

$$X2 = (W2 * (X0 + X1) + 2)$$

$$X3 = (W3 * (X0 + X1 + X2) + 3) \quad X4 = AvgP(X3) \tag{3}$$

and the lower Residual model is defined in equation:

$$X'1 = (W'1 * X0 + 1)$$

$$X'2 = (W'2 * (X'0 + X'1) + 2)$$

$$X'3 = (W'3 * (X'0 + X'1 + X'2) + 3)$$

X'4 = AvgP(X'3) (4) Inspired by the Inception module, we use the parallelism feature of the Inception module so that the upper and lower residual models work in parallel and are ultimately connected. The first three lines in each equation show the con volution operation for the data. We then feed the outputs of the third convolutional layers, namely, X3andX3 to the aver age pooling layer and then apply the dropout method (Fassihi et al., 2011).

Pooling layers

Our model has one pooling layer that performs average pooling with a filter size of 2 and a stride of 2. The average pooling function is defined in equation: Xi = AvgP(Xi−1) (5) where Xi−1 is the input data from the previous convolutional layer and AvgP denotes the average pooling operation. At the final stage, we apply a dropout technique with a probability of 0.25 directly after the average pooling layer (Pinheiro & Collobert, 2014).

The output data are then input to the soft-max classifier for classification. Thus, we have no fully connected layers other than the soft-max classifier. The first convolutional layer contains 153 trainable parameters. Each of the second and third convolutional layers contains 1305 trainable parameters (Farabet et al., 2013). For the University of Pavia dataset and the Pavia Centre scene dataset, the last FC layer contains 4140 trainable parameters; however, in the Salinas dataset, the last FC layer contains 14,704 parameters, and in the Indian Pines dataset, it contains 8109 parameters. This discrepancy occurs because the University of Pavia and Pavia Centre scene datasets each have nine output classes, whereas the Salinas dataset has 16 classes, and Indian Pines has 8 output classes. Thus, the total number of trainable parameters is 8,208 in the University of Pavia and Pavia Centre scene datasets, 18,772 in the Salinas dataset and 10,872 in the Indian Pines dataset.

## 4. SPECIFICATION

### 4.1 Hardware Requirements

The below Hardware Specifications were used in both Server and Client machines when developing.

Processor: Intel(R) Core (TM) i3
Processor Speed: 3.06 GHz
RAM: 256 MB
Hard Disk Drive: 40 GB
CD-ROM Drive: Sony
Monitor: “14” inches
Keyboard: TVS Gold
Mouse: Logitech

### 4.2 Software Requirements MATLAB

MATLAB is a tool for numerical computation and visualization. The basic data element is a matrix, so if you need a program that manipulates array-based data it is generally fast to write and run in MATLAB.

#### Getting Started

MATLAB is available on department machines. You can also download MATLAB for your personal machine from http://software.caltech.edu. Type “matlab” at the UNIX prompt to start. This will open the MATLAB desktop, which includes interactive menus and windows in addition to the command window.

## Goal

In this lab we will become familiar with the MATLAB programming environment. You will learn the basics of MATLAB syntax and functionality. You will also write your own program to do simple manipulations to an image in MATLAB.

## Description

MATLAB is a mathematical computer package from The Math Works. "MATLAB" stands for "Matrix Laboratory". MATLAB's functions operate on matrices and vectors, as well as scalar numbers, so it is very useful for linear algebra applications. With it, you can perform any number of simple or complicated mathematical operations, including image processing. As a general-purpose programming language, it is easy to learn and use, but powerful enough for a wide range of science and engineering applications. The best way to learn MATLAB is by using it, which is what we will be doing in this lab and throughout the semester. The built-in and online documentation is excellent and there is also a complete set of manuals for you to use. Please keep the manuals in the lab. To access MATLAB, log in to one of the PCs and double-click on the MATLAB icon on the desktop. The MATLAB environment will open. It is a window with several panes. The largest one is the command line. The others allow you to view the current directory, see what variables are in memory, etc. The menu bar at the top and the "Start" menu at the bottom of the window allow you to access many of MATLAB's functions as well as help and tutorials. **MATLAB** (**mat**rix **lab**oratory) is a multi-paradigmnumerical computing environment. A proprietary programming language developed by MathWorks, MATLAB allows matrix manipulations, plotting of functions and data, implementation of algorithms, creation of user interfaces, and interfacing with programs written in other languages, including C, C++, C#, Java, Fortran and Python. Although MATLAB is intended primarily for numerical computing, an optional toolbox uses the MuPADsymbolic engine, allowing access to symbolic computing abilities. An additional package, Simulink, adds graphical multi-domain simulation and model-based design for dynamic and embedded systems.As of 2017, MATLAB has over 2 million users across industry and academia. MATLAB users come from various backgrounds of engineering, science, and economics.

## Interfacing With Other Languages

MATLAB can call functions and subroutines written in the programming languages C or Fortran. A wrapper function is created allowing MATLAB data types to be passed and returned. The dynamically loadable object files created by compiling such functions are termed "MEX-files" (for **MA**TLAB **ex**ecutable). Since 2014 increasing two-way interfacing with Python is being added. Libraries written in Perl, Java, ActiveX or .NET can be directly called from MATLAB, and many MATLAB libraries (for example XML or SQL support) are implemented as wrappers around Java or ActiveX libraries. Calling MATLAB from Java is more complicated, but can be done with a MATLAB toolbox [33] which is sold separately by MathWorks, or using an undocumented mechanism called JMI (Java-to-MATLAB Interface), (which should not be confused with the unrelated Java Metadata Interface that is also called JMI). Official MATLAB API for Java was added in 2016. As alternatives to the MuPAD based Symbolic Math Toolbox available from MathWorks, MATLAB can be connected to Maple or Mathematical. Libraries also exist to import and export MathML.

### License

MATLAB is a proprietary product of MathWorks, so users are subject to vendor lock- in. Although MATLAB Builder products can deploy MATLAB functions as library files which can be used with .NET or Java application building environment, future development will still be tied to the MATLAB language. Each toolbox is purchased separately. If an evaluation license is requested, the MathWorks sales department requires detailed information about the project for which MATLAB is to be evaluated. If granted (which it often is), the evaluation license is valid for two to four weeks. A student version of MATLAB is available as is a home-use license for MATLAB, Simulink, and a subset of Mathwork's Toolboxes at substantially reduced prices. It has been reported that European Union (EU) competition regulators are investigating whether MathWorks refused to sell licenses to a competitor [43]. The regulators dropped the investigation after the complainant withdrew their accusation and no evidence of wrongdoing was found.

## 5. SYSTEM IMPLEMENTATION

### 5.1. Preprocessing

Preprocessing serves as the foundational step in preparing the input data for effective classification. In the context of dermo-cancer classification, preprocessing techniques play a crucial role in ensuring that the input skin images are standardized and devoid of any noise or artifacts that could interfere with accurate classification. Tasks such as resizing images to a uniform size, normalizing pixel values, and potentially augmenting the dataset through techniques like rotation, flipping, and scaling are commonly employed (LeCun et al., 1998). By standardizing the input data, preprocessing facilitates the extraction of relevant features and improves the robustness of the subsequent classification model.

## 5.2. Feature Extraction

Feature extraction is a pivotal aspect of automated dermo-cancer classification, wherein the model identifies and extracts distinctive characteristics from the input skin images. Techniques such as edge detection, texture analysis, and colour segmentation are utilized to capture relevant features indicative of dermo-cancer lesions (Havaei et al., 2014). In the proposed Hybrid Architecture, the integration of Inception and ResNet modules contributes significantly to feature extraction by leveraging their respective strengths in capturing multi-scale features and addressing challenges such as vanishing gradients. This hierarchical feature extraction process enables the model to discern intricate patterns in skin lesions, enhancing its ability to accurately classify dermo-cancer.

## 5.3. Data Augmentation

Data augmentation techniques play a crucial role in enhancing the robustness of the classification model by artificially expanding the training dataset. Through transformations such as rotation, flipping, and scaling, data augmentation introduces variability into the training data, thereby mitigating overfitting and improving the model's generalization ability.

By exposing the model to a diverse range of augmented images, data augmentation helps capture a broader spectrum of variations in dermo-cancer lesions, ultimately improving the model's performance on unseen data (Brosch et al., 2015).

## 5.4. Regularization

Regularization techniques are essential for preventing overfitting and promoting the generalization ability of the classification model. In the context of deep learning architectures like the proposed Hybrid Architecture, dropout regularization is commonly employed to mitigate the risk of over-reliance on specific features or neurons (Zikic et al., 2014). By randomly dropping a fraction of neurons during training, dropout encourages the network to learn more robust representations and reduces the likelihood of memorizing noise or irrelevant details present in the training data. This regularization mechanism helps improve the model's resilience to noise and enhances its performance on unseen data.

## 5.5. Model Evaluation and Validation

Model evaluation and validation are critical steps in assessing the performance and reliability of the proposed Hybrid Architecture. Metrics such as accuracy, sensitivity, specificity, and the area under the receiver operating characteristic curve (AUC-ROC) provide quantitative measures of the model's ability to correctly classify dermo-cancer lesions. Cross- validation techniques, such as k-fold validation, further enhance the reliability of performance metrics by evaluating the model on multiple subsets of the dataset (Urban et al., 2014). Through rigorous evaluation and validation procedures, the proposed system ensures the robustness and effectiveness of automated dermo-cancer classification.

## RESULT AND DISCUSSION

For this work, we collected a dataset of 800 images from website which were categorized into three stages. The dataset was divided into a training set of 80 images and a testing set of 20 images. We then trained the Hybrid model on the training set using the transfer learning approach, where the pre-trained weights of the Hybrid model were used as the initial weights for training. We fine-tuned the Hybrid model on the training set for 50 epochs, with a batch size of 10 and a learning rate of 0.0001.

A comprehensive collection of labeled data points utilized to train machine learning algorithms within the MATLAB environment.

*Figure 6. Trained data set*

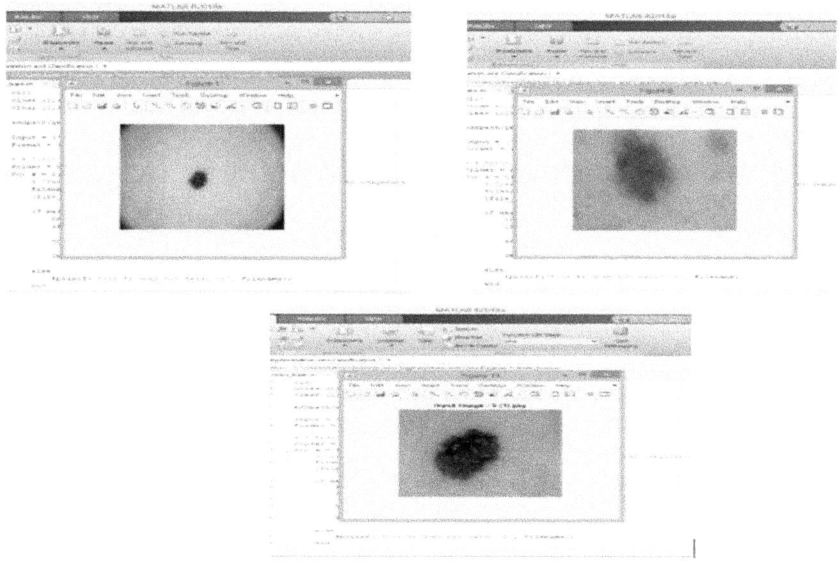

A visual input representing a specific scene or object, often used for testing and analysis in image processing tasks.

*Figure 7. Input test image*

An image that has undergone various transformations or enhancements, such as resizing, noise reduction, or feature extraction, to prepare it for further analysis or processing.

*Figure 8. Preprocessed image*

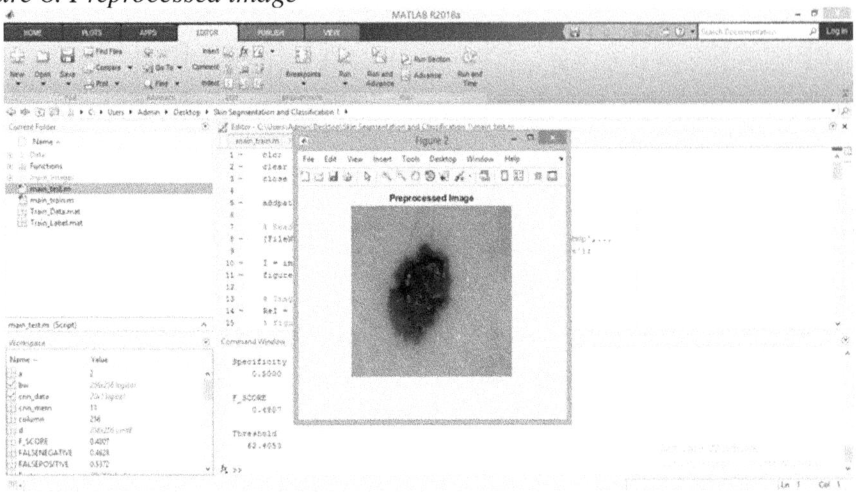


An image where each pixel is represented by a single binary value (typically 0 or 1), often used to depict objects or features in a scene as either present or absent.

*Figure 9. Binary image*

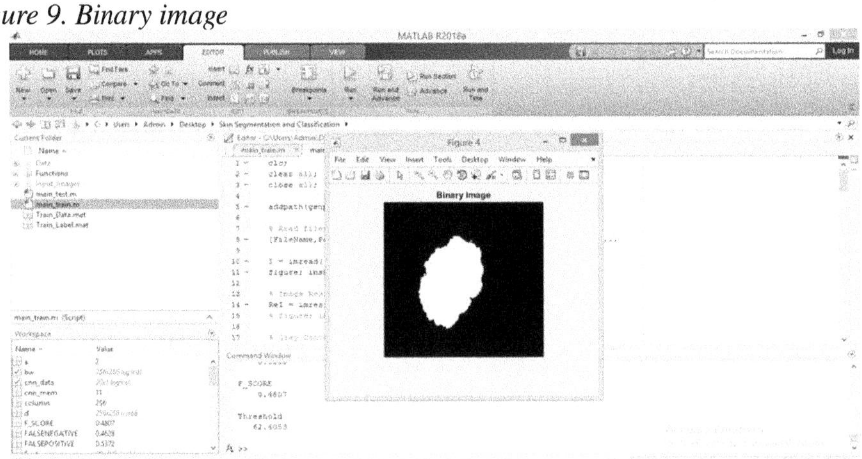

An image where regions of interest have been delineated using graph cut algorithms, separating them based on similarities in pixel intensity or other features, commonly employed in image segmentation tasks for object recognition or extraction.

*Figure 10. Segmented image by graph cut*

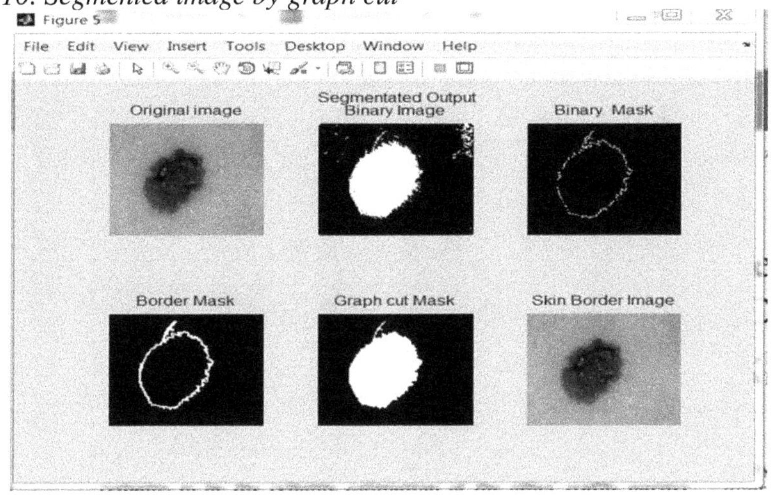


An analysis of similarities and differences between an input image and a dataset used for training, typically performed to assess how well the trained model generalizes to new or unseen data.

*Figure 11. Comparing input image with trained data set*

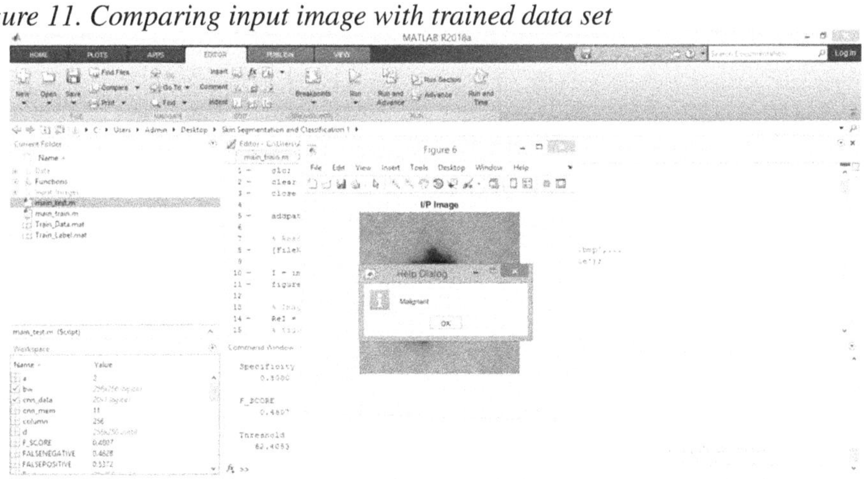

Table Performance Analysis of Precision, F1 Score, Accuracy and Specificity of the proposed method with Various Models

*Figure 12. Performance analysis*

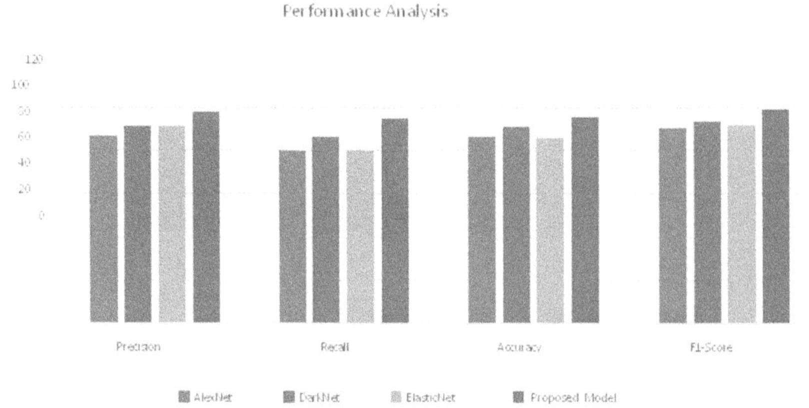


It summarizes the overall Precision, F1 Score Accuracy and Specificity values achieved by a proposed model in various Models.

*Table 1. Final Result*

| Models | Precision | Recall | Accuracy | F1-Score |
|---|---|---|---|---|
| GoogleNet | 86.51 | 79.45 | 86.22 | 90.33 |
| InceptionNet | 91.33 | 86.324 | 90.78 | 93.12 |
| ResNet | 91.32 | 80.12 | 85.6 | 91.75 |
| Proposed Model | 97.76 | 94.78 | 95.1 | 98.7 |

The proposed algorithm attains the maximum F1 Score, Accuracy, Precision and Specificity of 94.76, 94.84, 97.61 and 92.47 respectively.

## CONCLUSION

In conclusion, the integration of the Inception and ResNet modules within our proposed Hybrid Architecture presents a powerful solution for automated dermo-cancer classification from skin images. By leveraging the strengths of both architectures, we have developed a system capable of robust feature extraction and hierarchical learning, thereby enhancing the accuracy and efficacy of dermo-cancer classification. The Inception module excels at capturing multi-scale features and discerning intricate patterns within skin lesions, while the ResNet module addresses challenges such as vanishing gradients and facilitates the training of deeper neural networks. Through their fusion, our Hybrid Architecture achieves synergistic effects, combining the complementary features of both models to improve classification performance.

Experimental results on comprehensive datasets demonstrate the superiority of our approach compared to individual models, with enhanced accuracy, sensitivity, and specificity in dermo- cancer classification tasks. This underscores the effectiveness of hybrid architectures in medical image analysis and highlights the potential for advancing automated dermo-cancer diagnosis. In addition to its contributions to dermo-cancer classification, our proposed Hybrid Architecture holds promise for other medical image analysis tasks, showcasing the versatility and applicability of hybrid architectures in healthcare. By pushing the boundaries of automated classification systems, we aim to facilitate early detection and intervention for improved patient outcomes in dermatology and beyond.

## REFERENCES


Abuzaghleh, O.. (2015). Noninvasive real-time automated skin lesion analysis system for melanoma early detection and prevention. *IEEE Journal of Translational Engineering in Health and Medicine*, *3*, 1–12. https://doi.org/10.1109/JTEHM.2015.2435517

Aitken, J. F., Pfitzner, J., & Battistutta, D.. (1996). Reliability of computer image analysis of pigmented skin lesions of Australian adolescents. *Journal of Cancer*, *78*(2), 252–257. https://doi.org/10.1002/(SICI)1097-0215(19961009)78:2<252::AID-IJC10>3.0.CO;2-K

Amaral, T., Silva, L. M., & Alexandre, L. A.. (2014). Transfer learning using rotated image data to improve deep neural network performance. *International Conference on Image Analysis and Recognition (ICIAR: Image Analysis and Recognition)*, 290–300. https://doi.org/10.1007/978-3-319-11758-4_31

Aydinalp, C.. (2019). In vitro dielectric properties of rat skin tissue for microwave skin cancer detection. *IEEE Access : Practical Innovations, Open Solutions*, *7*, 115208–115216. https://doi.org/10.1109/ACCESS.2019.2935634

Badrinarayanan, V., Handa, A., & Cipolla, R. (2015). A deep convolutional encoder-decoder architecture for robust semantic pixel-wise labelling. *arXiv preprint arXiv:1505.07293*. https://doi.org/10.48550/arXiv.1505.07293

Behara, K., Bhero, E., & Agee, J. T. (2024). An improved skin lesion classification using a hybrid approach with active contour snake model and lightweight attention-guided capsule networks. *Diagnostics (Basel)*, *14*(6), 636. https://doi.org/10.3390/diagnostics14060636

Brosch, T., Yoo, Y., & Tang, L.. (2015). Deep convolutional encoder networks for multiple sclerosis lesion segmentation. *International Conference on Medical Image Computing and Computer-Assisted Intervention – MICCAI*, Vol. 9351, 3–11. https://doi.org/10.1007/978-3-319-24574-4_1

Bumrungkun, P.. (2018). Detection skin cancer using SVM and snake model. *2018 International Conference on Digital Arts, Media and Technology (ICDAMT)*, 46–49. https://doi.org/10.1109/ICDAMT.2018.8376521

C. K., Prasad, M. L., Kiran, A., Shaker Reddy, P. C., Babu, G. C., & Partheeban, N. (2023). Skin cancer classification using a hybrid convolutional neural network with SVM classifier. *2023 Global Conference on Information Technologies and Communications (GCITC)*, 1–6. https://doi.org/10.1109/GCITC60406.2023.10426054

Cancer Council Australia. (2010). Cancer Council to launch new research/failure to monitor highlights cancer risk. SunSmart/ Skin Cancer Facts and Figures. http://www.cancer.org.au/cancersmartlifestyle/SunSmart/Skincancer-factsandfigures.htm

Chang, Y., Stanley, R. J., & Moss, R. H.. (2005). A systematic heuristic approach for feature selection for melanoma discrimination using clinical images. *Skin Research and Technology : Official Journal of International Society for Bioengineering and the Skin (ISBS) and International Society for Digital Imaging of Skin (ISDIS) and International Society for Skin Imaging (ISSI)*, *11*(3), 165–178. https://doi.org/10.1111/j.1600-0846.2005.00133.x

Do, T. T.. (2015). Accessible melanoma detection using smartphones and mobile image analysis. *IEEE Transactions on Computational Imaging*, *1*(3), 176–184. https://doi.org/10.1109/TCI.2015.2476455

Farabet, C., Couprie, C., & Najman, L.. (2013). Learning hierarchical features for scene labeling. *IEEE Transactions on Pattern Analysis and Machine Intelligence*, *35*(8), 1915–1929. DOI: 10.1109/TPAMI.2012.231

Fassihi, N., Shanbehzadeh, J., & Sarrafzadeh, A.. (2011). Melanoma diagnosis by the use of wavelet analysis based on morphological operators. *Proceedings of the International Multiconference of Engineers and Computer Scientists*, 16–18.

Green, A., & Martin, N. (1994). Computer image analysis in the diagnosis of melanoma. *Journal of the American Academy of Dermatology*, *31*(6), 958–964. https://doi.org/10.1016/S0190-9622(94)70237-0

Havaei, M., Davy, A., & Warde-Farley, D.. (2014). Brain tumor segmentation with deep neural networks. *Proceedings of BRATS-MICCAI*.

Houssein, E. H., Abdelkareem, D. A., & Hu, G. (2024). An effective multiclass skin cancer classification approach based on deep convolutional neural network. *Cluster Computing*. Advance online publication. https://doi.org/10.1007/s10586-024-04540-1

Huang, J., & Jain, V. (2013). Deep and wide multiscale recursive networks for robust image labeling. *arXiv preprint arXiv:1310.0354*. https://doi.org/10.48550/arXiv.1310.0354

Kang, K., & Wang, X. (2014). Fully convolutional neural networks for crowd segmentation. *arXiv preprint arXiv:1411.4464*. https://doi.org/10.48550/arXiv.1411.4464

Keshavarz, A.. (2018). Water-based terahertz metamaterial for skin cancer detection application. *IEEE Transactions on Terahertz Science and Technology*, *8*(4), 397–405. https://doi.org/10.1109/TTHZ.2018.2839273

Kharazmi, P.. (2017). Automated detection and segmentation of vascular structures of skin lesions seen in dermoscopy, with an application to basal cell carcinoma classification. *IEEE Journal of Biomedical and Health Informatics*, *21*(6), 1675–1684. https://doi.org/10.1109/JBHI.2017.2676879

Korotkov, K.. (2015). A new total body scanning system for automatic change detection in multiple pigmented skin lesions. *IEEE Transactions on Medical Imaging*, *34*(1), 317–325. https://doi.org/10.1109/TMI.2014.2359991

LeCun, B., Bottou, L., & Bengio, Y.. (1998). Gradient-based learning applied to document recognition. *Proceedings of the IEEE*, *86*(11), 2278–2324. https://doi.org/10.1109/5.726791

Lee, H. C. (1994). *Skin cancer diagnosis using hierarchical neural networks and fuzzy logic*. Department of Computer Science, University of Missouri.

Long, E. S., Shelhamer, E., & Darrell, T. (2015). Fully convolutional networks for semantic segmentation. *Proceedings of the IEEE Conference on Computer Vision and Pattern Recognition*, 3431–3440. https://doi.org/10.1109/CVPR.2015.7298965

Long, J., Shelhamer, E., & Darrell, T. (2015). Fully convolutional networks for semantic segmentation. *IEEE Conference on Computer Vision and Pattern Recognition*, 3431–3440. https://doi.org/10.1109/CVPR.2015.7298965

Mirbeik, A.. (2018). Ultrawideband, stable normal and cancer skin tissue phantoms for millimeter-wave skin cancer imaging. *IEEE Access : Practical Innovations, Open Solutions*, *10*, 1109. https://doi.org/10.1109/ACCESS.2018.2805312

Mirbeik, A.. (2018). Synthetic ultra-high-resolution millimeter-wave imaging for skin cancer detection. *IEEE Transactions on Biomedical Circuits and Systems*, *12*(5), 1051–1061. https://doi.org/10.1109/TBCAS.2018.2858412

Pinheiro, P., & Collobert, R. (2014). Recurrent convolutional neural networks for scene labeling. *Proceedings of the 31st International Conference on Machine Learning*, 82–90.

Ronneberger, O., Fischer, P., & Brox, T. (2015). U-net: Convolutional networks for biomedical image segmentation. *Proceedings of the 18th International Conference on Medical Image Computing and Computer-Assisted Intervention (MICCAI)*, 8. https://doi.org/10.1007/978-3-319-24574-4_28

She, Z., Liu, Y., & Damatoa, A. (2007). Combination of features from skin pattern and ABCD analysis for lesion classification. *Skin Research and Technology : Official Journal of International Society for Bioengineering and the Skin (ISBS) and International Society for Digital Imaging of Skin (ISDIS) and International Society for Skin Imaging (ISSI)*, *13*(1), 25–33. https://doi.org/10.1111/j.1600-0846.2007.00177.x

Singaravelan, S., Selvakumar, V., Balaganesh, S., Gopalsamy, P., & Arun, R. (2024). Multiple feature fusion in particle filter framework for visual tracking. In *Improving Security, Privacy, and Connectivity Among Telemedicine Platforms* (pp. 238–258). IGI Global. https://doi.org/10.4018/978-1-6684-5958-1.ch010

Urban, B., Bendszus, M., & Hamprecht, F. A.. (2014). Multi-modal brain tumor segmentation using deep convolutional neural networks. *Proceedings of BRATS-MICCAI.*

Zikic, I., Ioannou, Y., & Criminisi, A.. (2014). Segmentation of brain tumor tissues with convolutional neural networks. *Proceedings of BRATS-MICCAI.*

Chapter 3
# Classification of Ovarian Cancer Subtype Using Histopathology Images

**R. Parvathi**
*Vellore Institute of Technology, Chennai, India*

**Xiaohui Yuan**
https://orcid.org/0000-0001-6897-4563
*University of North Texas, India*

## ABSTRACT

*Ovarian cancer is a common and reoccurring type of cancer. If detected early and treated properly, one can have a longer lifespan. We aim to provide robust alternative to the pre-existing system of manual diagnosis by automating the process using Deep Learning. In this paper we explore the three methods to classify histopathology images into its five corresponding types and outliers, possible cases of rare subtypes or healthy tissue misdiagnosed as cancerous. Previously, research made about detection and classification of other cancers have yielded significant results with several types of deep learning approaches, however the volume of data and annotations used for those is huge and difficult to obtain for all. Here, to make up for the lack of sufficient data, we have used transfer learning and fine tuning of pre-existing models trained for histopathological data.*

DOI: 10.4018/979-8-3693-3840-7.ch003

# 1 INTRODUCTION

Ovarian cancer affects almost 1 in every 87 women according to the American Cancer Society and every 1 in 130 has a probability of mortality (American Cancer Society, n.d.). The disease is asymptomatic at earlier stages and gets detected only in an advanced stage when it becomes difficult to treat and is often irresponsive to the commonly used chemotherapy. Examination of WSIs (Whole Slide Images) and TMAs (Tissue Microarray) is one of the ways in which this carcinoma can be classified into its subtypes.

Traditional methods of subtype classification based on manual histopathological assessment are labor-intensive, subjective, and prone to inter-observer variability. Machine learning based methods of CAD – which banks on the fact that the diagnostic errors made by humans and machines are fundamentally different and prevents overlooking by investigating all pixels, and CBIR – which works based on similarity in images and can accommodate weak and unlabeled data (Farahani et al., 2022) (Ibrahim et al., 2020).

While looking at these histopathological images the agreement between pathologists regarding the histotype is at best moderate (Cohen's kappa 0.54–0.67) (Farahani et al., 2022). There have been commendable results by using artificial intelligence to diagnose and histotype classification of breast cancer, lung cancer, and stomach cancer (Ibrahim et al., 2020)(Brunetti et al., 2022)(Ma et al., 2020). Using deep learning we aim to classify ovarian carcinomas into their histopathological subtypes to eliminate the confusion of categorizing manually.

In some previous works the various computer vision methods have yielded significant results in successfully categorizing the subtypes. However, these were done with much larger volumes of data or from a single source. Here, we shall deal with not only limited data and imbalanced data but also data collected from hospitals in 4 different competitions. This is the closest way to ensure the method produced can be applied universally.

Through empirical evaluations, this thesis aims to shed light on the utility of SSL and TL techniques as a viable solution for improving the accuracy and efficiency of ovarian cancer subtype classification. The findings from this research have the potential to impact clinical practice by facilitating more accurate diagnosis, personalized treatment planning, and prognostic assessment for patients with ovarian cancer. Furthermore, this work contributes to the broader field of digital pathology and machine learning by demonstrating the efficacy of machine learning in histopathological image analysis, with implications for other cancer types and biomedical applications.

## 2 LITERATURE SURVEY

Ovarian cancer presents a formidable challenge in accurate diagnosis and personalized treatment due to its diverse histopathological subtypes. In recent years, the intersection of machine learning techniques and histopathology image analysis has emerged as a promising avenue for subtype classification, offering improved diagnostic accuracy and treatment decision-making (Wang et al., 2024).

Efforts to classify ovarian carcinoma being classified using CT scans, genome testing, WSIs, MRIs, ultrasound, etc; have been made to enhance the efficiency of classification (Wang et al., 2024). As we know that the survival rate of cancer patients is highly correlated to it's early or late detection, it makes it important for us to develop a solution that can reliably diagnose and classify so that the corresponding treatment can be started immediately for best results.

The accurate categorization requires normalization of stain and lighting generate better results as they improve feature extraction and eliminate the need to manually re-stain (Michielli et al., 2022). Histopathology images have a huge size and it's important to segment them. Accurate segmentation of nuclei and glands in whole-slide images is essential for assessing abnormalities.

Data that is annotated in detail by field experts yields much better results however other methods can be used to boost the results obtained from data with no annotations or weak annotations (Van Eycke et al., 2019). Images can be transformed to mitigate overfitting by creating additional training samples through transformations. CNNs can be used to effectively manage the false positives and classify subtypes correctly within a range of 72-91% accuracy (Kasture et al., 2022) whereas ensemble learning of these models resulted in much better classification (Kassani et al., 2019). In applications with limited images, transfer learning works better, as it allows models to focus on relevant image regions and uses attention maps to visualize where the model looks during classification (Breen et al., 2023).

Overall, a combination of self-supervised leaning and transfer learning seem suitable approached to obtain firm results for limited and non-annotated data. Fine-tuned EfficientNet and .MIL can accurately predict instance labels. They can also leverage robust hierarchical pooling of features for bag-level predictions.

One such approach, self-supervised learning (SSL), extracts meaningful representations from unlabeled data, sidestepping extensive manual annotation. Pretext task learning and contrastive learning empower models to capture essential features crucial for accurate subtype classification of ovarian cancer (Wu et al., 2018)(Chen et al., 2020)(Doersch et al., 2015). Transfer learning further fortifies this endeavor by adapting pre-trained models to new tasks, thereby enhancing classification accuracy and efficiency in histopathology image analysis (Xu et al., 2019) (Yosinski et al., 2014).

Multiple Instance Learning (MIL) addresses the inherent challenges of weakly labeled datasets common in medical imaging. By organizing training data into bags containing multiple instances, MIL enables bag-level classification, suitable for subtype classification of ovarian cancer (Dietterich et al., 1997) (Wu et al., 2018). MIL effectively learns discriminative patterns from entire histopathology images, promising accurate classification even with limited annotation (Lazard et al., 2022).

Transformers and autoencoders offer complementary strategies by extracting hierarchical features and compact representations from histopathology images. These architectures capture long-range dependencies and latent representations, enriching the pool of features crucial for subtype classification (Ahmed et al., 2021). Moreover, neural networks play a pivotal role by automatically learning discriminative features, essential for differentiating between cancer subtypes based on learned representations (LeCun et al., 2015).

A notable advancement using iBOTv, combines self-supervised learning with transformer-based frameworks, automating cancer diagnosis and treatment planning. Leveraging unlabelled data and innovative techniques like Masked Image Modelling (MIM), iBOTv represents a significant leap in computational pathology (Filiot et al., 2023)(Tan & Le, 2019).

Further bolstering the arsenal of techniques is EfficientNet, optimizing efficiency and accuracy in subtype classification of ovarian cancer (Tan & Le, 2019) (Zelisse et al., 2024). By fine-tuning pre-trained models on histopathology images, EfficientNet extracts discriminative features, enhancing classification accuracy while optimizing computational resources (Hou et al., 2021) (Handley et al., 2022).

Attention-based Deep Multiple Instance Learning (ADMIL) integrates attention mechanisms with deep learning and MIL, achieving accurate and reliable subtype classification (Ilse et al., 2018) (Ahn et al., 2024). By dynamically weighing informative regions within histopathology images, ADMIL promises improved cancer diagnosis and treatment planning (Yao et al., 2020) (Farahani et al., 2022).

In conclusion, despite challenges such as dataset biases and model interpretability, the interdisciplinary nature of computational pathology holds promise for better outcomes in ovarian cancer care, driving towards more personalized and effective treatment strategies for ovarian cancer patients.

## 1.1. Challenges of Work

### 1.1.1. Complexity of Histopathological Images

Histopathology involves the study of diagnosis and study of diseases by looking at tissues under a microscope. The images produced are huge in size containing millions of pixels at a time (Doersch et al., 2015). Handling these images and operating on them requires a powerful processor and preferably a GPU and larger RAM.

They exhibit intricate patterns, cell structures, and tissue anomalies that demand professional attention. Tissue samples in histopathology can vary significantly in appearance based on factors such as tissue type, disease state, and individual variation among patients. This variability makes it challenging to develop automated algorithms or to interpret images consistently. Despite efforts to standardize diagnostic criteria, there can still be subjectivity in the interpretation of histopathology images. Different pathologists may interpret the same image differently, leading to variability in diagnoses.

### 1.1.2. Availablity of Labeled Data

Agreed upon by other papers (Xu et al., 2019), classification of images often relies on the goodness of its labels and extensiveness of its annotations. In contrast with natural images, high level annotations are not usually associated to particular objects in the image. In histopathological images, annotations are related to pathological lesions, morphological and architectural features, which encompass a complex mixture of visual patterns which allow us to decide about the illness presence, type and grade. In general, images with the same annotations present a high visual variability, which can be due to several factors, starting with the inevitable uncertainty coming from the fact that a very complicated 3D biological structure is projected onto a 2D image (Yosinski et al., 2014).

Expert-annotated data yields better results, but it is often scarce and expensive to obtain. Alternative methods, such as data augmentation and weak annotations, can enhance outcomes.

### 1.1.3. Normalization of Staining and Lighting

Achieving consistent staining and lighting conditions across histopathology images is crucial for accurate feature extraction. A stain too deep or light can hide important diagnostic details, reducing confidence, increasing the time for analysis, and possibly causing misdiagnosis. If deemed to be improper, the strain must be redone. However, proper normalization improves results and eliminates the need for manual re-staining (Dietterich et al., 1997).

Normalizing stains and lighting allow for more meaningful comparisons between different tissue samples and studies. For pathologists who visually inspect histopathology images to make diagnostic or prognostic assessments, normalized staining facilitates clearer visualization of tissue structures and abnormalities. This can improve the accuracy and consistency of diagnoses, leading to better patient outcomes.

In the case of multi-centre studies where histopathology images are collected from different institutions, these techniques enable data harmonization across sites, ensuring that findings are not confounded by variations in staining practices between different laboratories or hospitals.

### 1.1.4. Segmentation of Histopathology Images

Accurate segmentation of histopathology images helps identify the regions of interest. These ROIs can then be focused upon for quantification of various histological features, such as tumour size, shape, density, and distribution (Wu et al., 2018). Looking closely at parts of nuclei and glands within whole-slide images is essential for assessing abnormalities.

Automated segmentation methods are necessary for efficient analysis of such large images. Such segmented data can then be fed to ML algorithms for identification of features.

### 1.1.5. False Positive Management

Ovarian cancer subtypes have distinct prognostic implications, with some subtypes associated with better or worse outcomes compared to others. Misclassifying a patient's subtype as a false positive may lead to inaccurate prognostic assessments, potentially affecting treatment planning and patient counselling. It can result in patients receiving suboptimal or unnecessary treatments, leading to potential harm or reduced treatment efficacy.

False positive classifications can also introduce noise and bias into research studies and clinical trials investigating ovarian cancer subtypes. This over a period of time can lead to erroneous conclusions about disease biology, treatment efficacy, and patient outcomes. Managing false positives effectively can also help reduce healthcare expenditures and optimize resource utilization.

### 1.1.6. Limited and/or Imbalanced Data

With limited or imbalanced data, biases can occur and cause misdiagnosis. A study suggests that the quantity and quality of data is directly related to the results of the classification (Lazard et al., 2022).

A balanced dataset ensures that machine learning models are trained on representative samples from each class or subtype of ovarian cancer. This helps prevent bias towards overrepresented classes and improves the generalization ability of the models to accurately classify unseen data.

The histopathology images themselves can exhibit significant variability in terms of tissue appearance, staining characteristics, and imaging conditions. A diverse and balanced dataset helps machine learning models learn robust features that are invariant to such variability, leading to more robust and reliable classification performance.

## 1.2. Problem Statement and Objectives

. Ovarian cancer is a heterogeneous disease comprising various histological subtypes, each with distinct clinical behaviours, treatment responses, and prognostic implications. Accurate classification of ovarian cancer subtypes based on histopathology images is crucial for guiding treatment decisions, predicting patient outcomes, and advancing our understanding of disease biology. However, achieving reliable subtype classification poses significant challenges, including variability in tissue appearance, imbalanced datasets, and the need for sophisticated computational tools.

By leveraging machine learning techniques and large, balanced datasets, this project aims to address the challenges associated with accurate ovarian cancer subtype classification from histopathology images. The development of robust classification models holds promise for improving patient care, advancing research, and ultimately reducing the burden of ovarian cancer worldwide.

The objectives of this project are as follows:

a. A state-of-the-art machine learning-based framework for accurate and reliable classification of ovarian cancer subtypes from histopathology images.
b. Enhanced understanding of the histological features and patterns associated with different ovarian cancer subtypes.
c. Resilience to variations in tissue quality and staining will be investigated to ensure robustness of the developed machine learning algorithms.
d. Development of classification models for ovarian cancer subtypes using histopathology images without the need for annotated data will be explored.

## 1.3. Scope of the Research

The project's scope encompasses various pivotal components, delineated into five main facets: data collection and curation, stain normalization and preprocessing, feature extraction and model development, evaluation and validation, and clinical translation and impact.

Initially, the project endeavours to assemble a multifaceted dataset comprising histopathology images representative of diverse ovarian cancer subtypes. This entails ensuring a balanced distribution of samples across all subtypes to mitigate potential class imbalance issues, alongside obtaining requisite ethical approvals to ensure compliance with pertinent regulations governing data usage.

Subsequently, emphasis is placed on implementing stain normalization techniques to standardize the visual appearance of histopathology images, thereby minimizing variability introduced by disparate staining protocols. Preprocessing procedures, including color normalization, image registration, and artifact removal, are executed to augment the dataset's quality and consistency.

The project proceeds to deploy advanced machine learning algorithms, notably deep learning architectures such as convolutional neural networks, to extract discriminative features from the histopathology images. This culminates in the development of classification models adept at accurately delineating between different ovarian cancer subtypes based on the extracted features.

Evaluation and validation represent critical stages wherein the performance of the developed models is rigorously scrutinized using established validation protocols. This entails cross-validation and independent testing on unseen datasets to assess model accuracy, sensitivity, specificity, and robustness vis-à-vis variations in staining, imaging conditions, and tissue characteristics. Comparative analyses with existing classification methods and benchmarks are conducted to underscore the superiority of the proposed approach.

Finally, the project pivots towards clinical translation and impact, aiming to integrate the developed classification models into pathology workflows. Usability testing and validation studies involving pathologists are conducted to gauge the models' utility and efficacy in facilitating ovarian cancer subtype classification. An evaluation of the potential impact of enhanced classification on treatment decisions, patient outcomes, and healthcare resource utilization is undertaken through both retrospective and prospective studies.

## 3 METHODOLOGY

### 3.1 Dataset

For the purpose of this project, we have used a free and open dataset containing 538 histopathology images of ovarian cancer (Bashashati et al., 2023) (Wang et al., 2022). The data has been sourced from 20 hospitals spread over 4 continents. The data consists of actual images and their thumbnails, as well as a file containing dimensions and label of each image identified by their image ids. Of these images 25 images are TMAs at a magnification of 40x and the rest are WSIs at a magnification of 20x. The labels present in the dataset are:

i. High-Grade Serous Carcinoma (HGSC):
   Characteristics: HGSC is the most common and aggressive type of ovarian cancer, often diagnosed at an advanced stage in older women.
   Treatment: Surgery followed by chemotherapy.
ii. Low-Grade Serous Carcinoma (LGSC):
   Characteristics: LGSC is less common and slower-growing, typically diagnosed at an earlier stage.
   Treatment: Surgery, sometimes with targeted therapies.
iii. Endometrioid Carcinoma (EC):
   Characteristics: EC resembles uterine lining tissue (endometrium), often diagnosed early with a better prognosis.
   Treatment: Surgery, with possible chemotherapy or hormonal therapy.
iv. Clear Cell Carcinoma (CC):
   Characteristics: CC is rare, often chemo-resistant, and associated with a poorer prognosis.
   Treatment: Surgery, possibly with chemotherapy or targeted therapies.
v. Mucinous Carcinoma (MC):
   Characteristics: MC is less common, usually diagnosed early and localized.
   Treatment: Surgery, with consideration for chemotherapy based on disease extent.

Understanding these ovarian cancer types is crucial for tailoring treatment plans and predicting outcomes. Figure 1 shows the distribution of labels about the histopathological subtype of ovarian carcinoma and distribution of type of images, WSI or TMA

*Figure 1. Distribution of labels about the histopathological subtype of ovarian carcinoma and on the right is the distribution of type of images, WSI or TMA*

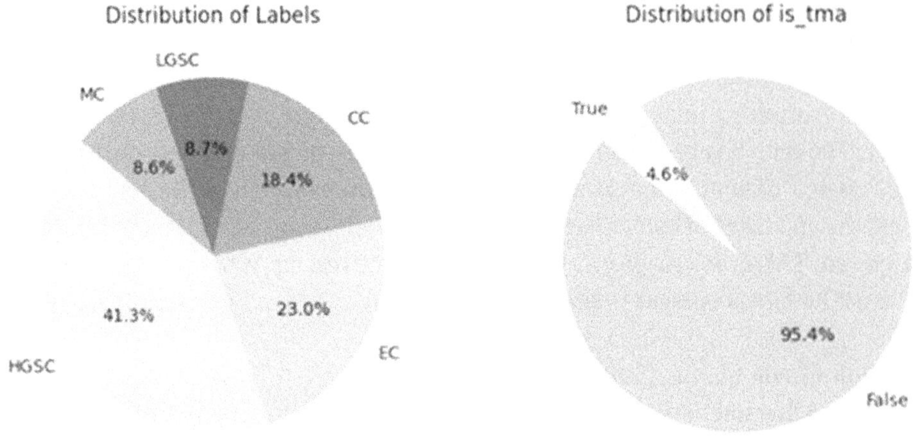


When working with large histology images, image pyramids play a crucial role in their classification. They help us analyze the structures present in the image that are often at different scales (Doersch et al., 2015). Image pyramids enhance the robustness of the classification model and aid in the acute localization of objects of interest, like tumor cells, dead cells and healthy cells.

## 3. 2 Models and Architecture

### 3.2.1 Convolutional Neural Network

We present a Convolutional Neural Network (CNN) architecture tailored for subtype classification, designed using the Keras Sequential API and its architecture shown in Figure 2.

*Figure 2. General architecture of classification pipeline*

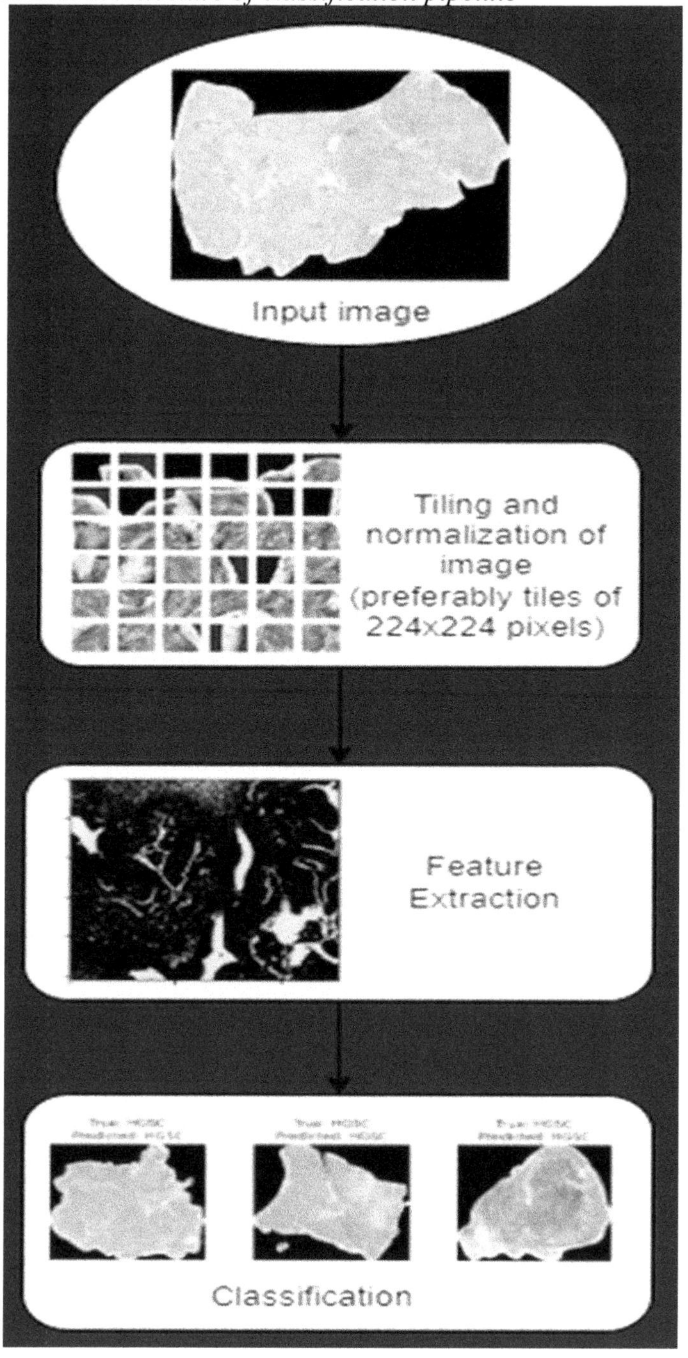


The CNN architecture consists of the following layers as shown in Figure 2:

i. Input Layer: The input layer accepts images of size (IMAGE_SIZE, IMAGE_SIZE, 3) where IMAGE_SIZE is the size of the input images and 3 represents the three colours channels (RGB).
ii. Convolutional Layers: The architecture includes multiple convolutional layers with ReLU activation functions. These layers extract features from the input images using learnable filters.
iii. Batch Normalization Layers: Batch normalization layers are added after each convolutional layer to stabilize and accelerate the training process by normalizing the activations.
iv. Max Pooling Layers: Max-pooling layers down sample the feature maps obtained from the convolutional layers, reducing the spatial dimensions while retaining important features.
v. Fully Connected Layers: The feature maps are flattened and passed through fully connected layers with ReLU activation functions. These layers enable the model to learn complex patterns and relationships in the data.
vi. Output Layer: The final layer is a dense layer with SoftMax activation, which outputs the predicted probabilities for each class.

*Figure 3. Sequential CNN architecture*

| Layer (type) | Output Shape | Param # |
|---|---|---|
| conv2d (Conv2D) | (None, 254, 254, 6) | 168 |
| batch_normalization (BatchNormalization) | (None, 254, 254, 6) | 24 |
| max_pooling2d (MaxPooling2D) | (None, 127, 127, 6) | 0 |
| conv2d_1 (Conv2D) | (None, 125, 125, 16) | 880 |
| batch_normalization_1 (BatchNormalization) | (None, 125, 125, 16) | 64 |
| max_pooling2d_1 (MaxPooling2D) | (None, 62, 62, 16) | 0 |
| flatten (Flatten) | (None, 61504) | 0 |
| dense (Dense) | (None, 8) | 492,040 |
| batch_normalization_2 (BatchNormalization) | (None, 8) | 32 |
| dense_1 (Dense) | (None, 2) | 18 |
| batch_normalization_3 (BatchNormalization) | (None, 2) | 8 |
| dense_2 (Dense) | (None, 5) | 15 |

Training employs the Adam optimizer and sparse categorical cross-entropy loss function. A training dataset is used for optimization, while a separate validation dataset evaluates performance during training. Early stopping prevents overfitting by monitoring performance on a testing dataset, halting training when improvement stalls or degradation begins.

The model is trained using the Adam optimizer with a sparse categorical cross-entropy loss function. We use a training dataset for model optimization and a separate validation dataset for evaluating the model's performance during training. The training process involves iterating over multiple epochs, with each epoch consisting of forward and backward passes through the network.

Early stopping is utilized to prevent overfitting by monitoring the model's performance on a testing dataset by stopping the training process when the performance stops improving or starts deteriorating. Our model is initially run for 100 epochs however on adding early stopping, it stops after 41 epochs.

### 3.2.2. Fine-Tuned EfficientNet

EfficientNet emerges as a promising candidate for ovarian cancer subtype classification, owing to its adaptive architecture and transfer learning capabilities. Its adept feature extraction, complemented by regularization techniques, ensures robust performance even with limited labelled data. The scalability of EfficientNet facilitates tailoring to dataset complexities, optimizing computational resources without compromising accuracy. Additionally, its interpretability aids in comprehending classification decisions, pivotal for clinical validation, thereby enhancing the accuracy, reliability, and clinical relevance of automated medical imaging analyses for ovarian cancer subtypes.

Preceding model training, the dataset undergoes preprocessing steps to bolster performance and generalization. These steps encompass image resizing, colour augmentation, geometric transformations (e.g., rotation, scaling), and pixel value normalization, efficiently implemented via the Augmentations library, ensuring resilience to input data variations.

In our study, model architecture choice shown in Figure 3 assumes paramount importance in achieving high classification accuracy. Leveraging a transfer learning approach, we adopt pre-trained EfficientNet architectures, with modifications to the final layers to suit the classification task requirements.

*Figure 4. Classification group based on feature map*

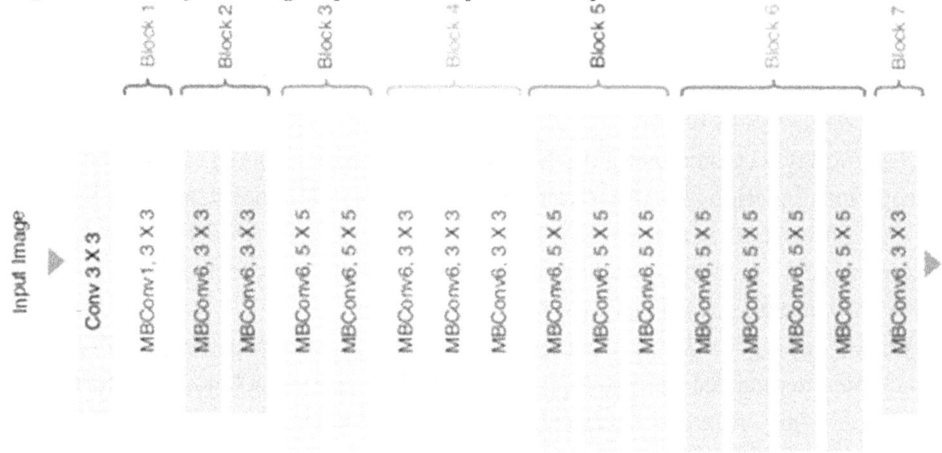


*Figure 5. Fine-tuned EfficientNet*

The customization is as follows as shown in Figure 4 and Figure 5:

- Efficientnetb0 base model: CNN-based backbone,
- pre-trained on imagenet
- final layers excluded
- Globalaveragepooling2d layer: reduce spatial dimensions and averages feature maps.
- Dense layer (1024 units, relu): fully connected
- adds non-linearity to the model and increases its capacity to learn complex patterns by introducing more parameters.
- Dense layer (5 units, softmax): final classification layer
- converts the raw scores into class probabilities, allowing the model to predict the class with the highest probability.
- Custom model compilation:
- sparse_categorical_crossentropy as the loss function since you have integer class labels.

- The optimizer used is Adam with a learning rate of 0.0001.
- Class weights calculation: handle class imbalance by giving more weight to underrepresented classes.
- Learning rate scheduler: custom learning rate scheduler function is defined to decrease the learning rate gradually after a few epochs to improve convergence.

The strategic integration of GeM (Generalized Mean Pooling) pooling arises from its efficacy in capturing salient features crucial for medical image classification tasks like ovarian cancer subtype classification. By computing a generalized mean over feature map spatial dimensions, GeM pooling accentuates informative regions while mitigating irrelevant areas, preserving spatial information vital for accurate classification. Its computational efficiency and parameter-free nature render it suitable for deployment in resource-constrained settings without compromising performance.

The training process iteratively optimizes model parameters to minimize a predefined loss function. Employing the Adam optimizer, chosen for its adaptive learning rate properties and efficient convergence, dynamic adjustments to the learning rate are facilitated through scheduling techniques like Cosine Annealing, facilitating effective navigation of the optimization landscape.

### 3.2.3 Fine-Tuned Phikon Feature Extractor With MIL Classifier

To effectively classify ovarian cancer subtypes, we employed a synergistic approach combining Phikon for feature extraction and Multiple Instance Learning (MIL) algorithms for classification. Motivated by MIL's robustness in handling the complexities of Whole Slide Image (WSI) analysis, our methodology aimed for comprehensive subtype classification.

We utilized Phikon, a sophisticated pre-trained network specifically tailored for WSI image description, as our feature extractor. Trained on a diverse set of 17 datasets, Phikon ensured a comprehensive understanding of histological images, leveraging Vision Transformer (ViT) models as its backbone. Through multitask learning, including Masked Image Modeling and Weakly-Supervised Classification tasks, Phikon generated embeddings encapsulating essential features relevant to ovarian cancer subtype classification.

For classification, we employed MIL algorithms, recognized for its effectiveness in handling features from multiple WSI image segments. Each WSI image was divided into patches, constituting bags in the MIL framework. The MIL classifier, facilitated the transition from local descriptors to global descriptors, enabling effective aggregation of patch-level information as given in the Figure 7.

In our context, each bag corresponds to the patches extracted from an individual WSI image. Chowder, implemented as a one-dimensional convolutional neural network (CNN), served as our primary MIL algorithm. This architecture facilitated the transition from local descriptors to global descriptors, enabling effective aggregation of information from the diverse set of patches within each image as shown Figure 6.

### 3.1.1. Custom Model of Phikon and Admil

*Figure 6. Phikon network training structure*

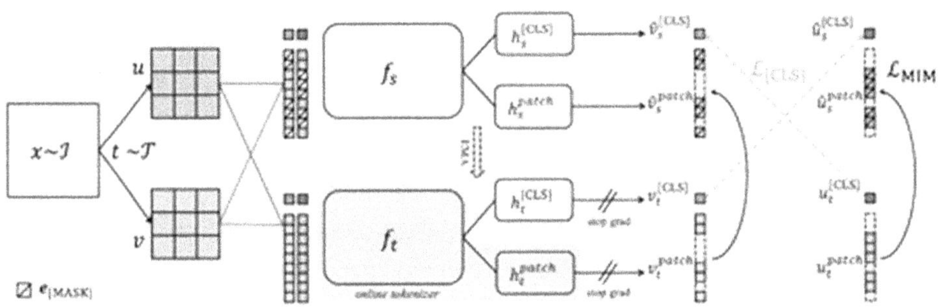

*Figure 7. Fine-tuned Phikon feature extractor with MIL Classifier*

*Figure 8. Phikon feature extractor with Chowder classifier*

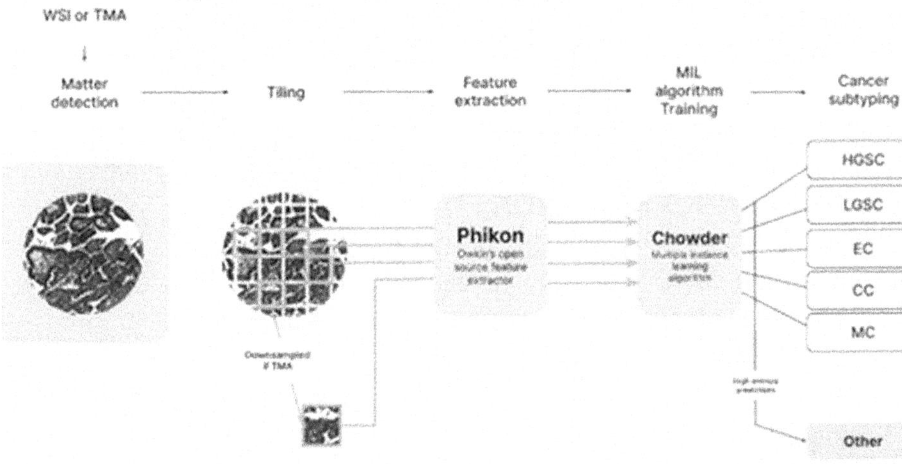


Our model architecture comprised three main components as shown in Figure 7:

i. Image Preprocessing:

  - Identification of regions of interest through thresholding over the Hue-Saturation-Value (HSV) image.
  - Tiling of images into patches, with varying patch sizes depending on the image type (WSI or TMA).

ii. Feature Extraction with Phikon:

  - Phikon network extracts embeddings characterizing each patch, leveraging its comprehensive understanding of histological images.

iii. Attention-based Deep Multiple Instance Learning (MIL) Classification:

  - Chowder algorithm processes the embeddings from individual patches, aggregates them into a single representation, and performs classification.

This holistic approach amalgamates advanced feature extraction capabilities with robust classification methodologies, resulting in a high-performance classifier adept at discerning between ovarian cancer subtypes. By integrating Phikon and MIL algorithms, we contribute to enhanced diagnostic accuracy and improved patient outcomes in clinical settings as shown in the Figure 8.

Additionally, our architecture includes specialized modules:

1. DimReduction Module: Reducing input feature dimensionality using linear transformation followed by ReLU activation, optionally incorporating residual blocks.
2. Attention_Gated Module: Incorporating attention mechanisms to focus on relevant features, producing attention weights to emphasize important features.

## 4 RESULTS AND DISCUSSION

The evaluation of various convolutional neural network (CNN) architectures, including a fine-tuned EfficientNet and a fine-tuned Phikon feature extractor with a Multiple Instance Learning (MIL) classifier, was conducted to ascertain their efficacy in classifying ovarian cancer subtypes. Figure 9 shows the Accuracy vs Epoch for training and validation set using CNN and Figure 10 shows the Loss vs Epoch for training and validation set using CNN.

*Figure 9. Accuracy vs Epoch for training and validation set using CNN*

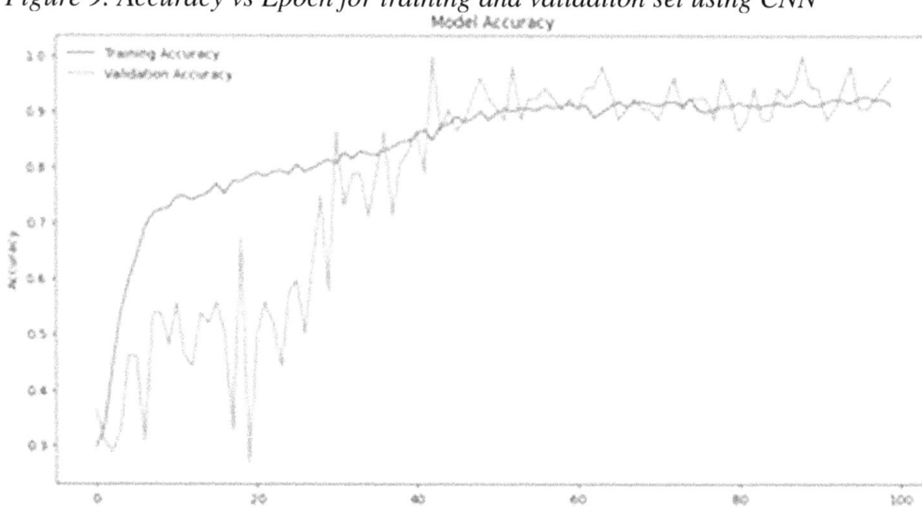

*Figure 10. Loss vs Epoch for training and validation set using CNN*

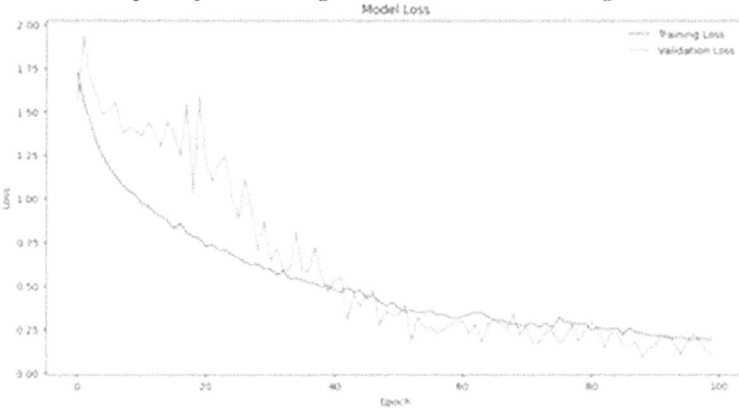

The CNN architecture demonstrated commendable performance, achieving an accuracy of 87.62% on the training set and 86.54% on the validation set. However, analysis of accuracy versus epoch and loss versus epoch graphs revealed signs of overfitting to the validation set, indicating potential limitations in generalization capability as shown in Figure 11and Figure 12.

*Figure 11. Accuracy vs Epoch for training and validation set using EfficientNet*

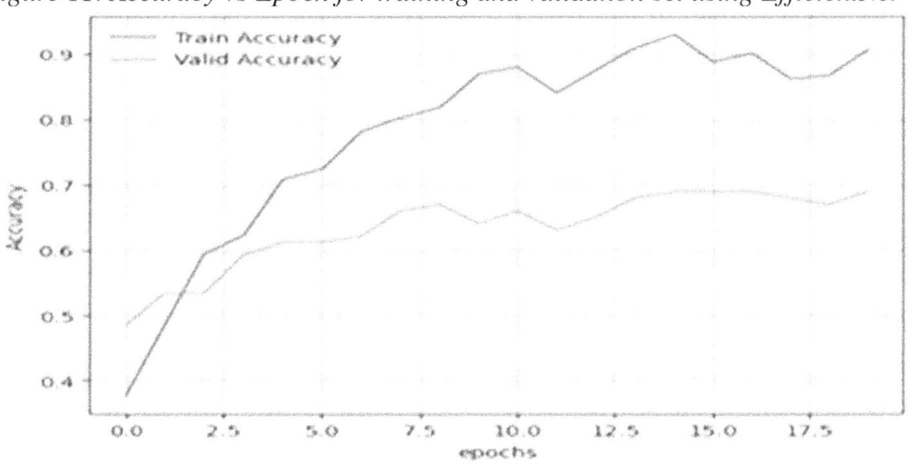

*Figure 12. Loss vs Epoch for training and validation set using EfficientNet*

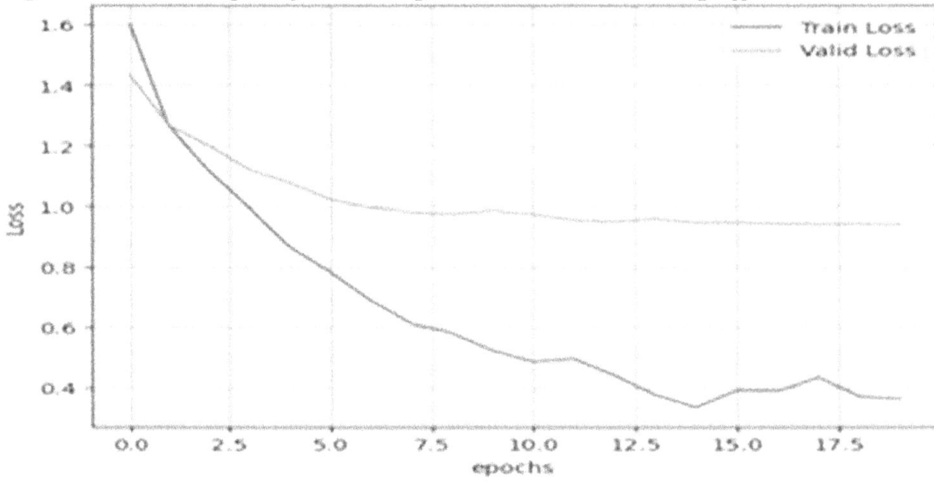


In contrast, the fine-tuned EfficientNet exhibited a training accuracy of 90.6%, yet its validation accuracy was notably lower at 68.9%. This discrepancy may be attributed to the absence of pre-training on histopathological data, leading to potential overfitting to the training dataset. Despite this limitation, the EfficientNet model provided valuable insights into feature extraction and classification performance as shown in Table 1.

*Table 1. Class-wise accuracy of MIL model*

| **Class** | **Accuracy** |
|---|---|
| CC | 96.96% |
| EC | 89.51% |
| HGSC | 85.49% |
| LGSC | 93.61% |
| MUC | 86.96% |

Remarkably, the fine-tuned Phikon feature extractor with MIL classifier yielded the highest accuracy among the evaluated models, achieving an impressive accuracy of 93.49%. This substantial performance improvement can be attributed to the pre-training of Phikon on a diverse range of histopathological data, enabling it to effectively capture relevant features for subtype classification.

*Figure 13. Confusion matrix for model*

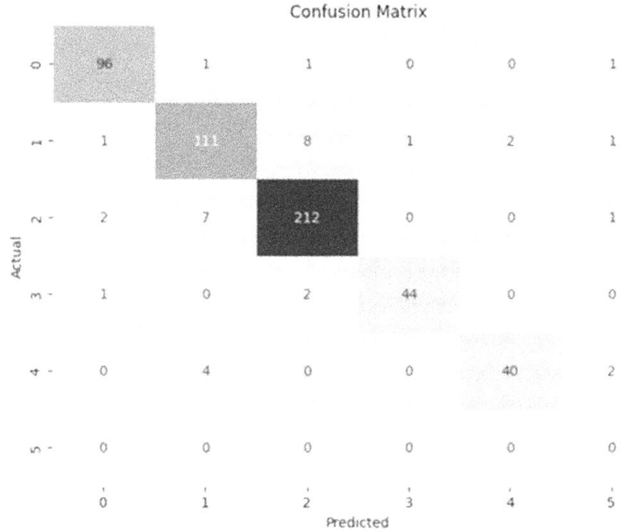


Figure 13 depicts the Confusion matrix the actual and predicted values of different classes like High-Grade Serous Carcinoma (HGSC), Low-Grade Serous Carcinoma (LGSC), Endometrioid Carcinoma (EC), Clear Cell Carcinoma (CC), Mucinous Carcinoma (MC)

Additionally, the implementation of early stopping mechanisms across all models mitigated the risk of overfitting and ensured the robustness of the selected models. The selection of the best-performing model based on validation accuracy further enhanced the reliability of the results.

Furthermore, the comprehensive evaluation of model performance included the examination of confusion matrices and class-wise accuracies, providing a detailed understanding of classification outcomes across different ovarian cancer subtypes as shown in Table 2.

*Table 2. Comparison of accuracies of models across labels*

| **Model name** | **Average accuracy** |
|---|---|
| CNN | 87.62% |
| EfficientNet | 90.6% |
| Phikon and Chowder | 93.49% |

Comparative analysis revealed that the model utilizing the Phikon feature extractor with the MIL classifier outperformed both the baseline CNN architecture. CNN model provided 87.62, EfficientNet has given 90.6 and Phikon with Chowder architecture gave 93.49% of accuracy as summarized in Table 2.

These findings underscore the significance of transfer learning approaches, particularly when leveraging pre-trained models specifically tailored for histopathological image analysis, such as Phikon. The integration of advanced feature extraction techniques with robust classification methodologies holds promise for enhancing the accuracy and clinical relevance of automated ovarian cancer subtype classification systems.

Overall, this study contributes to the ongoing efforts aimed at improving diagnostic accuracy and patient outcomes in the realm of histopathological image analysis, laying the groundwork for future research endeavors in this domain.

## 5 CONCLUSION

Histopathological images pose challenges due to their complexity and variability. Obtaining labeled data is crucial for classification, but it's scarce and costly. Normalizing staining and lighting ensures consistent visibility, while segmentation helps identify regions of interest. Managing false positives is critical for accurate assessments. Balanced datasets improve model generalization, and diverse data enhances performance. In summary, addressing complexity, ensuring data quality, and using transfer learning are vital for accurate ovarian cancer classification.

The successful application of our proposed methodology highlights its potential for clinical translation, with significant implications for patient outcomes and treatment decision-making. Accurate classification of ovarian cancer subtypes from histological images allows tailored treatment approaches based on individual tumor profiles. Additionally, automated diagnostic tools can revolutionize clinical practice by facilitating rapid and accurate subtype classification while reducing the burden on healthcare providers.

Looking ahead, further research in outlier detection and model refinement holds promise for enhancing the classifier's adaptability and scalability. Integrating complementary data sources, such as genomic and clinical data, could provide valuable insights into underlying biological mechanisms driving ovarian cancer subtypes. Our findings underscore the transformative potential of deep learning in oncology, offering personalized and effective treatments for ovarian cancer patients.

Collaboration between data scientists, clinicians, and researchers is essential to harnessing AI's full potential in improving cancer diagnosis and prognosis. Future work should explore larger datasets and the integration of annotated and unannotated data. Leveraging semi-supervised or self-supervised learning techniques could improve generalization performance and robustness.

Alternative Multiple Instance Learning (MIL) techniques, attention mechanisms, and ensemble modeling offer exciting avenues for enhancing feature representation and diagnostic accuracy. The development of an automated ovarian cancer subtype classifier can streamline clinical workflows, but it should always complement expert judgment. Ultimately, our work contributes to advancing personalized treatments and patient care.

## REFERENCES


Ahmed, S., Shaikh, A., Alshahrani, H., Alghamdi, A., Alrizq, M., Baber, J., & Bakhtyar, M. (2021, August 9). Transfer Learning Approach for Classification of Histopathology Whole Slide Images. *Sensors (Basel)*, *21*(16), 5361. DOI: 10.3390/s21165361 PMID: 34450802

Ahn, B., Moon, D., Kim, H. S., Lee, C., Cho, N. H., Choi, H.-K., Kim, D., Lee, J.-Y., Nam, E. J., Won, D., An, H. J., Kwon, S. Y., Shin, S.-J., Jung, H. R., Kwon, D., Park, H., Kim, M., Cha, Y. J., Park, H., & Park, E. (2024). Histopathologic image–based deep learning classifier for predicting platinum-based treatment responses in high-grade serous ovarian cancer. *Nature Communications*, *15*(1), 4253. DOI: 10.1038/s41467-024-48667-6 PMID: 38762636

American Cancer Society – ovarian cancer key statistics: https://www.cancer.org/content/dam/CRC/PDF/Public/8773.00.pdf

Bashashati, A. (2023). *UBC ovarian cancer subtype classification and outlier detection*. UBCOCEAN.

Big Self-Supervised Models are Strong Semi-Supervised Learners, Chen, Ting and Kornblith, Simon and Swersky, Kevin and Norouzi, Mohammad and Hinton, Geoffrey, arXiv preprint arXiv:2006.10029, 2020

Breen, J., Allen, K., Zucker, K., Hall, G., Orsi, N., & Ravikumar, N. (2023). Efficient subtyping of ovarian cancer histopathology whole slide images using active sampling in multiple instance learning. 10.48550/arXiv.2302.08867.

Brunetti, A., Altini, N., Buongiorno, D., Garolla, E., Corallo, F., Gravina, M., Bevilacqua, V., & Prencipe, B. (2022). A Machine Learning and Radiomics Approach in Lung Cancer for Predicting Histological Subtype. *Applied Sciences (Basel, Switzerland)*, *12*(12), 5829. DOI: 10.3390/app12125829

Dietterich, T. G., Lathrop, R. H., & Lozano-Pérez, T. (1997). Solving the multiple instance problem with axis-parallel rectangles. *Artificial Intelligence*, *89*(1-2), 31–71.

Doersch, C., Gupta, A., & Efros, A. A. (2015). Unsupervised visual representation learning by context prediction. In *Proceedings of the IEEE international conference on computer vision* (pp. 1422-1430).

Farahani, H., Boschman, J., Farnell, D., Darbandsari, A., Zhang, A., Ahmadvand, P., & Bashashati, A. (2022). Deep learning-based histotype diagnosis of ovarian carcinoma whole-slide pathology images. *Modern Pathology*, *35*(12), 1983–1990.

Farahani, H., Boschman, J., Farnell, D., Darbandsari, A., Zhang, A., Ahmadvand, P., Jones, S. J. M., Huntsman, D., Köbel, M., Gilks, C. B., Singh, N., & Bashashati, A. (2022). Deep learning-based histotype diagnosis of ovarian carcinoma whole-slide pathology images. *Modern Pathology*, *35*(12), 1983–1990. DOI: 10.1038/s41379-022-01146-z PMID: 36065012

Filiot, A., Ghermi, R., Olivier, A., Jacob, P., Fidon, L., Mac Kain, A., Saillard, C., & Schiratti, J.-B. Scaling Self-Supervised Learning for Histopathology with Masked Image Modeling. *medRxiv* 2023.07.21.23292757; DOI: 10.1101/2023.07.21.23292757

Handley, K. F., Sims, T. T., Bateman, N. W., Glassman, D., Foster, K. I., Lee, S., Yao, J., Yao, H., Fellman, B. M., Liu, J., Lu, Z., Conrads, K. A., Hood, B. L., Barakat, W., Zhao, L., Zhang, J., Westin, S. N., Celestino, J., Rangel, K. M., & Sood, A. K. (2022). Classification of High-Grade Serous Ovarian Cancer Using Tumor Morphologic Characteristics. *JAMA Network Open*, *5*(10), e2236626. DOI: 10.1001/jamanetworkopen.2022.36626 PMID: 36239936

Hou, Q., Zhou, D., & Feng, J. (2021). Coordinate attention for efficient mobile network design. In *Proceedings of the IEEE/CVF conference on computer vision and pattern recognition* (pp. 13713-13722).

Ibrahim, A., Gamble, P., Jaroensri, R., Abdelsamea, M. M., Mermel, C. H., Chen, P. H. C., & Rakha, E. A. (2020). Artificial intelligence in digital breast pathology: Techniques and applications. *The Breast : Official Journal of the European Society of Mastology*, *49*, 267–273.

Kassani, S. H., Kassani, P. H., Wesolowski, M. J., Schneider, K. A., & Deters, R. (2019). Classification of histopathological biopsy images using ensemble of deep learning networks. arXiv preprint arXiv:1909.11870.

Kasture, K. R., Choudhari, D., & Matte, P. N. (2022). Prediction and classification of ovarian cancer using enhanced deep convolutional neural network. *International Journal of Engineering Trends and Technology*, *70*(3), 310–318.

Khashaba, M., Fawzy, M., Abdel-Aziz, A., Eladawei, G., & Nagib, R. (2022). Subtyping of high grade serous ovarian carcinoma: Histopathological and immunohistochemical approach. *Journal of the Egyptian National Cancer Institute*, *34*(1), 6. DOI: 10.1186/s43046-022-00104-9 PMID: 35138498

Lazard, T., Bataillon, G., Naylor, P., Popova, T., Bidard, F. C., Stoppa-Lyonnet, D., Stern, M. H., Decencière, E., Walter, T., & Vincent-Salomon, A. (2022, December 20). Deep learning identifies morphological patterns of homologous recombination deficiency in luminal breast cancers from whole slide images. *Cell Reports Medicine*, *3*(12), 100872. DOI: 10.1016/j.xcrm.2022.100872 PMID: 36516847

LeCun, Y., Bengio, Y., & Hinton, G. (2015). Deep learning. *Nature*, *521*(7553), 436–444. DOI: 10.1038/nature14539 PMID: 26017442

Ma, B., Guo, Y., Hu, W., Yuan, F., Zhu, Z., Yu, Y., & Zou, H. (2020, October 2). Artificial Intelligence-Based Multiclass Classification of Benign or Malignant Mucosal Lesions of the Stomach. *Frontiers in Pharmacology*, *11*, 572372. DOI: 10.3389/fphar.2020.572372 PMID: 33132910

Michielli, N., Caputo, A., Scotto, M., Mogetta, A., Pennisi, O. A. M., Molinari, F., & Salvi, M. (2022). Stain normalization in digital pathology: Clinical multi-center evaluation of image quality. *Journal of Pathology Informatics*, *13*, 100145.

Tan, Mingxing & Le, Quoc. (2019). EfficientNet: Rethinking Model Scaling for Convolutional Neural Networks.

Van Eycke, Y. R., Foucart, A., & Decaestecker, C. (2019). Strategies to reduce the expert supervision required for deep learning-based segmentation of histopathological images. *Frontiers in medicine*, *6*, 222.

Wang, C. W., Chang, C. C., Khalil, M. A., Lin, Y.-J., Liou, Y.-A., Hsu, P.-C., Lee, Y.-C., Wang, C.-H., & Chao, T.-K. (2022). Histopathological whole slide image dataset for classification of treatment effectiveness to ovarian cancer. *Scientific Data*, *9*(1), 25. DOI: 10.1038/s41597-022-01127-6 PMID: 35087101

Wang, Y. L., Gao, S., Xiao, Q., Li, C., Grzegorzek, M., Zhang, Y. Y., Li, X. H., Kang, Y., Liu, F. H., Huang, D. H., Gong, T. T., & Wu, Q. J. (2024, March 11). Role of artificial intelligence in digital pathology for gynecological cancers. *Computational and Structural Biotechnology Journal*, *24*, 205–212. DOI: 10.1016/j.csbj.2024.03.007 PMID: 38510535

Wu, Z., Xiong, Y., Yu, S. X., & Lin, D. (2018). Unsupervised Feature Learning via Non-parametric Instance Discrimination. *2018 IEEE/CVF Conference on Computer Vision and Pattern Recognition*, 3733-3742. DOI: 10.1109/CVPR.2018.00393

Wu, Z., Xiong, Y., Yu, S. X., & Lin, D. (2018). Unsupervised feature learning via non-parametric instance discrimination. In *Proceedings of the IEEE conference on computer vision and pattern recognition* (pp. 3733-3742).

Xu, Y., Hosny, A., Zeleznik, R., Parmar, C., Coroller, T., Franco, I., Mak, R. H., & Aerts, H. J. W. L. (2019, June 1). Deep Learning Predicts Lung Cancer Treatment Response from Serial Medical Imaging. *Clinical Cancer Research*, *25*(11), 3266–3275. DOI: 10.1158/1078-0432.CCR-18-2495 PMID: 31010833

Yao, J., Zhu, X., Jonnagaddala, J., Hawkins, N., & Huang, J. (2020). Whole slide images based cancer survival prediction using attention guided deep multiple instance learning networks. *Medical Image Analysis*, *65*, 101789.

Yosinski, J., Clune, J., Bengio, Y., & Lipson, H. (2014). How transferable are features in deep neural networks? *Advances in Neural Information Processing Systems*, •••, 27.

Zelisse, H. S., Hwan, R. A., van de Vijver, M. J., Dijk, F., Mom, C. H., Hooijer, G. K. J., van Gent, M. D. J. M., & Snijders, M. L. H. (2024). Evaluation of the prognostic potential of histopathological subtyping in high-grade serous ovarian carcinoma. *Virchows Archiv*. Advance online publication. DOI: 10.1007/s00428-024-03807-7 PMID: 38622323

# Chapter 4
# Computer-Aided Algorithms for Underwater Image Enhancement and Object Detection

**Naveena T. Joseph**
https://orcid.org/0009-0000-7532-1341
*Amal Jyothi College of Engineering, India*

**S. N. Kumar**
https://orcid.org/0000-0002-2530-1454
*Amal Jyothi College of Engineering, India*

## ABSTRACT

*The underwater world holds a wealth of secrets as it dominates three fourth of the earth surface. However, capturing underwater images with clear details is not an easy task. Light scattering, absorption, and limited visibility create a challenging environment for traditional cameras. The various phenomena that degrade underwater images and how it affect image clarity are examined. The image enhancement techniques, both traditional methods and cutting-edge deep learning approaches, which tackle issues like color correction, contrast enhancement, and haze removal, effectively restoring the true colors and details hidden beneath the water's surface are considered. Image enhancement plays a crucial role as a pre-processing step, paving the way for more accurate object detection. Various deep learning-based frameworks specifically designed to excel in these challenging conditions are explored in this chapter.*

DOI: 10.4018/979-8-3693-3840-7.ch004

# 1. INTRODUCTION

70% of the earth is filled with ocean. The world hidden below the ocean is a garden with mysteries and wonders. Capturing the beauty of this garden is challenging. On each attempt, the image captured is distorted because of light scattering resulting in an image with low contrast, color cast, and obscured details (Tarekegn et al., 2023). This chapter aims to describe computer-aided algorithms that can be applied to one underwater image covering 2 key areas namely, Underwater image enhancement and underwater object detection. Underwater image enhancement aims to recover the exact color and texture of the underwater world by incorporating color correction, gradient boosting, and image fusion algorithms (Hou et al., 2019). Underwater object detection pinpoints each object present within the image. These algorithms help to explore the marine environment, and coral reef structures, and monitor underwater resources (Li et al., 2023). Compared to traditional methods, the use of computer-aided algorithms provides promising results and ensures consistency. These methods aim at improving efficiency by automating image processing and object detection (Han et al., 2020). Several preprocessing methods like noise reduction, color correction, and white balance aid in achieving good outcomes.

Underwater images pose significant challenges when compared to an images taken on land. Such images are characterized by low contrast, blurred details, and color distortion. Contrast difference is caused mainly by the scattering and absorption of light in water. Enhancement of underwater images plays a crucial role in converting blurred images to a clear one. These methods help to remove the deep color tint using color correction and provide a natural image. Contrast enhancement methods can increase the sharpness and improve the edge details of the image making them more realistic (Woon, 2020). As a result, object detection becomes easier since the objects are distinct from their background, allowing researchers to locate

The visual capability of the human eye varies from person to person and those for an underwater image are too low. Sophisticated tools are needed in this field for better clarity and visualization of images. Based on this demand, research was carried out and algorithms were designed. These algorithms have helped marine biologists to conduct population surveys, species identification, and monitor the health and extent of coral reefs (Wang et al., 2023) (Gonzalez-Rivero et al., 2020). Underwater archeologists can use these tools for detecting submerged cultural heritage sites for detailed study (Kuzmin & Grekov, 2021). These algorithms can contribute to the search and rescue missions by locating objects and survivors with increased speed and accuracy. The bias and subjectivity that occur during manual processes can be eliminated with computer-aided algorithms, thereby generating more reliable results. Such algorithms can be easily deployed in real-time environments, helping autonomous underwater vehicles to make decisions in deep-sea projects.

## 2. FUNDAMENTALS OF UNDERWATER IMAGING

The quality and clarity of the captured underwater environment depend greatly on the properties of the water. Algorithms developed by considering these factors can produce better results. Light traveling through water encounters absorption, attenuation, and scattering and together contribute to the degradation of image quality. As the light penetrates deeper into the water column, the water molecules and dissolved particles absorb light resulting in reduced light intensity (Yang et al., 2023). Color distortion and loss of contrast is due to this wavelength-dependent absorption and hence shorter wavelengths are absorbed rapidly than longer wavelengths. Scattering causes light to strike suspended particles and molecules in water and hence deviate from its original path and scatter in multiple directions. This results in reduced visibility and increased background noise in underwater images. The size of the scattered particles, their concentration, and light wavelength are other factors that influence scattering.

Attenuation occurs due to decrease in light intensity as it travels down the water column. It also includes scattering and absorption, thereby leading to loss of light energy with increasing depth. This creates a barrier to the effective range of the underwater imaging system, facilitating the design of specialized imaging techniques to compensate for the reduced light levels.

*Figure 1. Model of underwater image acquisition*

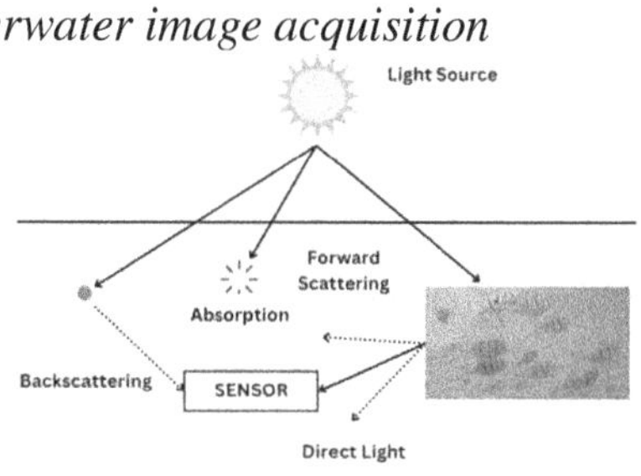


An extensive study was carried out by researchers on the optical properties of water and their impact on underwater imaging to derive new techniques for enhancing image quality while compensating for the effect of light attenuation and scattering. Mobley (1994) has done a detailed analysis of challenges and opportunities of underwater imaging considering the optical properties of water and its impact on remote sensing applications (Garcia et al., 2002). Jerlovand Atkinson conducted a detailed study on light absorption and scattering in aquatic environments which has served as a background for other researchers (Mobley, 1994). To capture and visualize the submerged environment, a wide variety of technologies were designed in underwater imaging systems (Jerlov, 1976). Each of these technologies has diverse sensing modalities with unique advantages and applications. SONAR, LiDAR, and optical imaging systems are the most unique ones.

**Sound Navigation and Ranging(SONAR):** SONAR imaging system creates an image of underwater terrain and structures by emitting soundwaves and capturing the echoes reflected from the objects. It is mainly used in areas with high turbidity and low visibility. For shallow water environments, side scan sonar imaging is effective (Jaffe, 1990). Studies by shows that emitted acoustic pulses to the sides of a sensor platform producing high-resolution images. A detailed 3-dimensional map of the underwater surface can be created by using multiple beams of acoustic energy. Based on research, hydrothermal vent fields and underwater volcanic features can be mapped by a multi-beam sonar system (Tritech International Limited, 2008).

**Light Detection And Ranging (LiDAR):** LiDAR systems generate high-resolution 3D maps of underwater environments by measuring distance using laser pulses (Medwin & Clay, 1998). This technology was recently deployed in underwater imaging, even though it was used in terrestrial applications. This system can be used for mapping coral reef and monitoring reef health was dictated by the study of Burns (He & Seet, 2002). Visible or infrared light is used to capture underwater images in the optical imaging system. For capturing still images or videos of underwater environments, underwater cameras are specially designed with waterproof housings and specialized technologies. A study was conducted by Makynen in 2017 using underwater cameras and depicted its use in monitoring fish populations and studying marine ecosystems (Burns et al., 2022). For capturing images with a wide range of wavelengths, hyperspectral imaging systems are used. This technology captures images that can be used for advanced spectral analysis and was used by Mobley in 2017 for water quality parameters and object detection investigation (Mäkynen & Karvonen, 2017). Integration of advanced image processing algorithms together with these technologies can deliver better results, which can aid researchers in having a better understanding and help them to research various challenging areas of marine science, exploration, and resource management.

## 3. UNDERWATER IMAGE ENHANCEMENT TECHNIQUES

Attenuation and scattering of the light causes noise and color correction in the captured underwater image, affecting its quality and clarity. Preprocessing techniques helps to improve its visual perception to a great extent. Several algorithms were developed and were renewed over years, based on the findings in image processing and computer vision. Filtering methods are used to remove noise from underwater images. Mean filter preserves edge details along with which it removes noise

(Mobley et al., 2004). In this, the intensity of each pixel is replaced with median of all neighboring pixels. Speckle noises in underwater images were removed by this technique in a research conducted by Zhan (2018) (Lu & Serikawa, 2014). Multi-resolution properties of wavelet transform are used to eliminate noise from image details in wavelet based denoising method. The entire image is decomposed into several parts based on frequency and selective denoising is applied separately on each of them. Studies by Yang (2019) demonstrated wavelet denoising to suppress noise and enhance image quality (Latif et al., 2018). Inorder to improve the contrast and brightness of underwater images, the pixel intensities are redistributed to cover the entire dynamic range in histogram equalization. This method helps to overcome uneven illumination, thereby enhancing image details (Wang et al., 2019).

In underwater images, absorption and scattering causes color distortion, which can be eliminated by color constancy algorithms. Color correction methods improve color fidelity and visual realism by compensating the effect of ambient light and water turbidity (Patel et al., 2020). More accurate color correction is done with physically based models that simulate optical properties of water. Factors such as water depth, light spectrum and absorption coefficients are considered to produce realistic color rendering, generating good performance compared to traditional methods. Dehazing methods work by removing the haze or fog, generating good contrast and fine detailed underwater scenes. This is mainly due to light attenuation and scattering. Single images mainly use Dark Channel Prior (DCP) for dehazing (Zhang et al., 2017). This method operates by assuming that in haze free region, the intensity of atleast one color channel is close to zero in natural images. This property is used to eliminate the depth of haze and recover a haze free image. Dehazing algorithms also considers the amount of atmospheric light scattered by particles in scene using statistical and physical model based approaches. This method improves performance by considering both water surface and underwater environment. Effectiveness of dehazing algorithms can be enhanced using multi-fusion techniques in which information from different scales are combined (Peng et al., 2018). The variation in scene depth and haze density can be handled by adding local and global features into dehazing methods. Color attenuation characteristics of underwater scenes are bring together for estimating and removing haze in Color Attenuation Prior(CAP) method (Zhang et al., 2023). This method distinguish haze from picture content by its color distribution, resulting in increased dehazing performance even in turbid underwater environment. Developments in the field of artificial intelligence and deep learning have led to the deployment of dehazing methods based on neural networks. Deep learning based underwater dehazing method uses Conventional Neural Network (CNN) to create a mapping between haze free and hazy image data, producing promising results compared to traditional method (Ngo et al., 2019).

Contrast enhancement algorithms are designed to reduce the contrast lost due to light attenuation and scattering in underwater scenario by distinguishing an object from its background and henceforth attains increased visibility and quality. Histogram equalization is the most common contrast enhancement technique that makes the image more appropriate for interpretation by humans and machines (Gui et al., 2023). This method stretches the histogram of the image and higher contrast is obtained with greater stretching. In an extended work, histogram of the image is divided into small regions and histogram equalization is applied on each region independently. Adaptive histogram equalization entitles that the contrast enhancement is based on the local image characteristics (Kaur et al., 2011). Inorder to prevent over enhancement of contrast and removal of artifacts, the enhancement of contrast is limited in areas with large intensity variations. This approach of Contrast Limited Adaptive Histogram Equalization (CLAHE) preserves the overall image quality by applying amplification to selected regions (Pizer et al., 1987). In retinex-based algorithms, illuminance and reflectance components of an image are separated to restore local contrast of the image. Illumination component is only enhanced preserving the image details. Another important algorithm mainly used in underwater imagery for contrast enhancement is Multi-Scale Retinex with Color Restoration (MSRCR) (Zuiderveld, 1994). Wavelet transforms can be used in contrast enhancement to decompose the image into different frequency bands and each band is enhanced selectively. Wavelet based approach preserves the image details while enhancing the contrast across the bands (Parthasarathy & Sankaran, 2012).

## 4. OBJECT DETECTION IN UNDERWATER IMAGES

### 4.1 Challenges in Underwater Object Detection

Complex nature of the underwater environment raises a unique set of challenges for object detection. Some are low visibility, environmental factors such as varying light condition, dynamic state of water, and variation in object appearance. Suspended particles in water, light attenuation and water turbidity causes low visibility of underwater images, which in turn affects the quality and range of sensor data, making the performance of object detection algorithms very low (Priyadharsini et al., 2018).

Water distortions, occulations and varying light conditions may cause underwater object to exhibit diverse appearances. This causes challenges in object detection since it orientation, distance from source and environmental factors makes their appearance different. State of underwater environment is dynamic in nature, due to turbulent waves and currents can add noise and motion blur to sensor data, resulting in undistinguishable objects from its background. Limited data availability is another factor that hinders object detection in underwater images. Training and testing of object detection algorithms are costly because of unavailability of annotated data. Benchmark dataset are to be established for evaluating the algorithm. Sensors used in the underwater environment plays a crucial role in obtaining input data and images. Resolution, range and sensitivity of the sensors are limited, which thereby impacts the accuracy and reliability of object detection. Selection of sensors and its conFigureuration plays a crucial role (Gomes et al., 2020).

## 4.2 Computer Aided Algorithms Used for Under Water Object Detection

Over the past few years, there was a drastic change in technology with regard to the accuracy in object detection. Traditional object detection frameworks are being replaced by deep learning based approaches, for locating and identifying the component objects in the image or video frame. Traditional methods uses multiple handcrafted features with different stages particularly, feature extraction, pattern recognition and classification. A feature descriptor based Histogram of Oriented Gradients(HOG) counts the occurrence of gradients orientation in localized portions of an image. It uses both magnitude and angle of orientation for feature computation. HOG along with SVM classifier is used in pedestrian detection and other applications (Er et al., 2023). A feature based haar cascade classifier that extracts haar features mainly for face detection applications. Based on these haar features images are classified as positive and negative (Dalal & Triggs, 2005).

Convolutional Neural Network (CNN) can automatically learn features directly from image without any handcrafted feature extraction (Shamrat et al., 2022). Deep learning frameworks use unified architectures that implements object localization and classification. R-CNN and its derivatives (Fast R-CNN, Faster R-CNN) identify Regions of Interest (RoIs) in an image through techniques like selective search, followed by classifying these regions using a CNN (Sung et al., 2017). Despite achieving remarkable accuracy, these approaches are computationally intensive. SSD stands for Single Shot Multibox Detector, which is an object detection framework capable of real-time performance. It predicts both object bounding boxes and class probabilities directly from feature maps across various scales (Girshick, 2015). SSD strikes a balance between accuracy and speed by employing a single feedforward CNN for object detection, YOLO, or You Only Look Once, is a real-time object detection framework that treats object detection as a regression task, predicting spatially bounded boxes and their corresponding class probabilities. YOLO analyzes the entire image in a single pass through the network, resulting in high efficiency (Kanimozhi et al., 2019).

Object detection algorithms designed specifically for underwater environments utilize a range of methods to tackle the distinct challenges presented by underwater imaging conditions, such as poor visibility, diverse object appearances, and environmental influences. Faster R-CNN, a widely-used object detection framework, has been modified for underwater use. It employs a region proposal network (RPN) to generate region proposals and a convolutional neural network (CNN) to detect objects within these proposals. Faster R-CNN is utilized for underwater object detection, showcasing its efficacy in identifying marine life and submerged structures (Sung et al., 2017). YOLO, another object detection framework, has been tailored for underwater scenario. It functions by dividing the input image into a grid and predicting bounding boxes and class probabilities for each grid cell. Research utilized YOLO for detecting underwater objects, attaining real-time performance and reliable accuracy in challenging, turbid underwater environments (Kanimozhi et al., 2019). SSD, a real-time object detection framework, functions by predicting bounding boxes and class probabilities across various scales within feature maps. SSD is expanded for detecting underwater objects, showcasing its capability to identify marine species and submerged objects accurately and efficiently. Mask R-CNN, an extension of Faster R-CNN, predicts segmentation masks alongside detecting objects. It not only identify objects but also separate them from the background (Terven & Cordova-Esparza, 2023). Mask R-CNN is utilized for underwater object detection and segmentation, attaining accurate localization and delineation of marine organisms and underwater structures. R-FCN is a region-based object detection framework that integrates fully convolutional networks (FCNs) to achieve both accuracy and efficiency (He et al., 2017). It has the capability to precisely localize objects. R-FCN is also utilized for detecting underwater objects, showing its effectiveness in identifying submerged objects in demanding underwater environments.

## 5. INTEGRATION OF UNDERWATER IMAGE ENHANCEMENT AND DETECTION ALGORITHMS

Combining enhancement and detection techniques is vital for enhancing performance in underwater imaging applications. Enhancement methods enhance underwater image quality by mitigating issues like light attenuation, scattering, and color distortion, while detection techniques locate and identify objects within the improved images. Environmental factors like poor visibility, fluctuating illumination, and color distortion are addressed by modifying image contrast, brightness, and color balance. The outcomes of object detection offer crucial insights to enhancement algorithms, directing them to prioritize enhancing regions where objects are expected. By incorporating detection feedback into the enhancement procedure, algorithms can concentrate on improving relevant image areas, resulting in more focused and efficient enhancement (Dai et al., 2016). This approach enables an end-to-end optimization, where both elements are collectively trained and refined to attain peak performance and enables the system to dynamically adjust enhancement parameters based on detection feedback, resulting in enhanced overall performance.

The process of integrating enhancement and detection algorithms comprises multiple steps focused on enhancing the quality of input data to enhance detection performance. The workflow begins with preprocessing in which noise reduction can be applied to diminish noise in the input data, employ methods like median filtering, wavelet denoising, or Gaussian smoothing. To boost image contrast, employ techniques such as histogram equalization, adaptive histogram equalization, or contrast limited adaptive histogram equalization (CLAHE) for contrast enhancement. Inorder to refine image details while retaining edges, employ methods like unsharp masking, bilateral filtering, or multiscale Retinex for detail enhancement. For analyzing texture, extract texture features utilizing techniques such as Gabor filters, local binary patterns (LBP), or gray-level co-occurrence matrices (GLCM). To identify prominent edges within an image, utilize edge detection algorithms like the Canny edge detector, Sobel operator, or Prewitt operator. To build a classifier or detector, train it using features derived from the improved images. Techniques such as Support Vector Machines (SVM), Random Forests, or Convolutional Neural Networks (CNNs) can be utilized for this purpose. If traditional methods are used for object detection, employ conventional detection algorithms like template matching, Hough transform, or clustering techniques. As post processing steps, implement methods such as non-maximum suppression, morphological operations, or post-classification filtering to refine detection results and mitigate false positives or negatives. Assess the effectiveness of the integrated enhancement and detection algorithms by measuring metrics like precision, recall, F1-score, receiver operating characteristic (ROC) curve, or area under the curve (AUC).

With the aim of monitoring and protecting marine diversity, an automated image enhancement and classification pipeline is developed, which uses the images taken by Wally – an Internet Operated Vehicle (IOV). Residual Convolutional Neural Network is used to enhance the acquired images whereas classification uses CNN and deep neural networks, generating better results (Wei et al., 2020). For accurate detection of Chinese Mitten crab and to classify its gender, a light weight network based on improved YOLOv4, GMNet–YOLOv4 is developed. GhostNet is used as the backbone of YOLOv4 for crab feature extraction. Depth wise separable convolution replaces ordinary convolution, thereby increasing accuracy and reducing parameters (Lopez-Vazquez et al., 2023). Underwater search and rescue operation faces challenges in detecting small targets in turbid and complex underwater environment. In sonar based acoustic imaging, an improved YOLOv5 along with ShuffleNetv2 is designed with improved accuracy. A hierarchical transfer learning based two-branch convolution module enhances feature extraction, making it a light weight model for easy deployment on embedded systems. Experimental results show an improved mean average precision (mAP) compared to traditional detection algorithms and also reduces computational requirements (Chen et al., 2023).

For underwater biological detection, Self-Adaptive Global Histogram Stretching (SAGHS) and Convolutional Block Attention Module (CBAM) are integrated into YOLOv5 network. SAGHS includes a self-adaptive contrast correction and a self-adaptive color correction module. Self-adaptive contrast correction module decomposes the RGB channels and color equalization is applied. For self-adaptive color correction, RGB image is converted to CIELAB color space and a linear histogram stretching is applied on the L component and is converted back to RGB. Channel attention (CA) and spatial attention (SA) mechanisms are used to detect features of tiny marine organisms, extract and conserve them. CA exploits the inter channel relationship of features whereas SA make use of internal spatial relationship of features. This method demonstrates increased performance in image quality and accuracy of detection, proving its applicability in real-world applications like ecological monitoring and evaluation of marine resources (Hu & Liu, 2024)

## 6. EVALUATION METRICS AND BENCHMARKING

Metrics are quantifiable measures for evaluating a process. Both enhancement and detection process are to be evaluated separately and to main metric for each case are PSNR, SSIM and precision, recall, mAP.

PSNR (Peak Signal-to-Noise Ratio): PSNR evaluates the efficacy of an image enhancement algorithm by contrasting the original image with its enhanced counterpart. Elevated PSNR values signify greater resemblance between the original and enhanced images, indicating minimal distortion or information loss during enhancement (Liu et al., 2022). Conversely, a reduction in PSNR indicates heightened distortion or noise introduced during the enhancement process.

$$PSNR = 10 \log_{10} \frac{MAX^2}{MSE}$$

Where MAX is the maximum possible pixel value of the image and MSE is mean squared error.

SSIM (Structural Similarity Index): SSIM assesses the structural likeness between the original and enhanced images, factoring in luminance, contrast, and structure. Its scale ranges from -1 to 1, with 1 representing perfect similarity. Enhanced images with higher SSIM values retain more image structures and details (Liu et al., 2022).

Conversely, a decline in SSIM indicates a deterioration in image quality resulting from enhancement.

$$SSIM(x,y) = \frac{\left(2\mu_x\mu_y + C_1\right)\left(2\sigma_{xy} + C_2\right)}{\left(\mu_x^2 + \mu_y^2 + C_1\right)\left(\sigma_x^2 + \sigma_y^2 + C_2\right)}$$

Where $\mu_x$, $\mu_y$ are mean intensity values of image x and y. $\sigma_x$ and $\sigma_y$ are standard variation.

Precision: Precision assesses the accuracy of object detection by determining the proportion of correctly identified objects among all objects detected (Liu et al., 2022). An uptick in precision indicates a reduction in false positives, indicating that the algorithm more precisely identifies relevant objects while minimizing inclusion of irrelevant ones.

$$Precision = \frac{TP}{TP + FP}$$

Where TP is True Positive and FP is False Positive.

Recall: Recall measures the comprehensiveness of object detection by determining the proportion of correctly identified objects relative to the total number of objects present in the scene. A rise in recall signifies a reduction in false negatives, indicating that the algorithm effectively detects all relevant objects within the scene (Dosselmann & Yang, 2006).

$$Recall = \frac{TP}{TP + FN}$$

Where TP is True Positive and FN is False Negative.

mAP (mean Average Precision): mAP calculates the average precision across all classes in object detection tasks. A higher mAP indicates superior overall performance in detecting objects across various classes, while a decrease reflects inferior performance (Buckland & Gey, 1994).

Benchmarking underwater image enhancement and object detection algorithms frequently demands datasets designed specifically to address the unique challenges posed by underwater environments. The Underwater Image Enhancement Benchmark (UIEB) dataset comprises a diverse array of underwater images taken under various conditions (Lin et al., 2014). It encompasses images exhibiting different degrees of degradation, including blurriness, color cast, and low contrast. An expanded version of the UIEB dataset, UIEB-Refined, includes manually refined ground truth images for improved evaluation accuracy. This dataset offers more precise annotations, enhancing the benchmarking process for image enhancement algorithms. The UIEB-

HAZE Dataset is centered on underwater haze removal. It comprises underwater images exhibiting different levels of haze, along with corresponding ground truth images for evaluation purposes (Lin et al., 2014).

The Underwater Detection Dataset (UWDD) is tailored for assessing object detection algorithms in underwater settings. It comprises annotated images featuring a variety of underwater objects, including fish, corals, and man-made structures (Li et al., 2020). The UWA3D Multispectral Dataset offers multispectral images depicting underwater scenes across diverse environments. It features annotations for various objects and marine life, making it well-suited for assessing multispectral object detection algorithms (Chen et al., 2020). The R/V Oceanus Dataset comprises images captured by underwater robotic vehicles across a range of marine environments. It encompasses annotations for various marine organisms and objects, aiding in the assessment of underwater object detection algorithms (Ren et al., 2021).

## 6.1 Challenges in Benchmarking Underwater Algorithms and Potential Solutions

Underwater environments experience fluctuations in image quality due to factors like lighting conditions, which are influenced by variables such as time of day, weather, and depth. For instance, images taken during daylight near the water's surface tend to be clearer and more vivid due to abundant natural light. However, at greater depths or in dark conditions, artificial or specialized underwater lighting may be required for adequate illumination.

Water clarity plays a vital role in determining image quality underwater. Clear, unpolluted water enables light to penetrate deeper, resulting in sharper and more detailed images. Conversely, murky or turbid water, caused by suspended particles like sediment or algae, can greatly diminish visibility and image clarity. This decrease in water clarity can lead to haziness or blurriness in images, posing challenges for imaging systems to accurately capture fine details.

Acquiring precise ground truth data for underwater images, particularly in natural environments, presents challenges due to factors such as water distortion and restricted visibility. The shortage of extensive and varied datasets tailored for evaluating underwater algorithms impedes both their development and assessment. Object detection and segmentation in underwater scenes pose significant challenges due to their complexity, characterized by diverse marine life, structures, and varying backgrounds. Algorithms trained on datasets from above-water environments may struggle to perform well in underwater scenarios due to domain shift, necessitating the use of domain adaptation techniques.

Creating standardized evaluation protocols can promote consistency in benchmarking endeavors by defining metrics, datasets, and evaluation procedures. Synthetic data generation methods enable the creation of extensive datasets featuring diverse underwater scenes, facilitating controlled variations and annotations. Enhancing current datasets through methods such as color manipulation, simulated blur, and noise incorporation can enrich the training and evaluation data, resulting in greater diversity. By employing transfer learning and domain adaptation techniques, such as fine-tuning pre-existing models or integrating domain-invariant features, algorithm performance in underwater environments can be enhanced. Promoting collaboration among researchers and organizations to exchange datasets, annotations, and evaluation methodologies can facilitate the advancement and benchmarking of underwater algorithms. Performing field trials and real-world testing across various underwater environments offers valuable insights into algorithm performance in practical scenarios. Combining data from diverse sensors like sonar, LiDAR, and multispectral imaging can bolster the resilience and precision of underwater algorithms.

## 7. APPLICATIONS AND FUTURE DIRECTIONS

Computer-assisted algorithms are pivotal in numerous facets of underwater exploration, marine biology, and underwater archaeology, empowering researchers to analyze data, uncover discoveries, and comprehend underwater ecosystems. In underwater exploration, computer-based algorithms play critical roles in areas such as Underwater Mapping and Surveying, Autonomous Underwater Vehicles (AUVs) and Search and Rescue Operations. In Underwater Mapping and Surveying, sonar, LiDAR, and photogrammetric data are explored to generate detailed maps of underwater terrain and features, facilitating exploration and navigation. In Autonomous Underwater Vehicles (AUVs), independent navigation, avoiding obstacles, optimizing routes, and conducting surveys of underwater areas for scientific research and resource exploration is implemented using these algorithms. In Search and Rescue Operations, Computer vision algorithms analyze underwater imagery from drones or remotely operated vehicles (ROVs) to identify and locate objects of interest.

In marine biology, species identification and tracking, habitat mapping and monitoring and bioacoustic analysis are main areas in which computer based algorithms are used. In species identification and tracking, these algorithms analyze underwater images to identify and track marine species, aiding researchers in population monitoring, behavioural studies, and ecosystem health assessments. Algorithms process data from underwater sensors and imaging systems to map marine habitats, monitor temporal changes, and evaluate the influence of environmental factors on marine ecosystems. Site discovery and mapping, artifact detection and classification and 3d reconstruction and visualization are the main areas covered in underwater archaeology. By analysing sonar and image data, algorithms detect and map underwater archaeological sites, such as shipwrecks, submerged cities, and ancient artifacts for further study. Image processing algorithms scrutinize underwater imagery to identify, categorize, and reconstruct archaeological artifacts, aiding in the documentation and preservation of underwater cultural heritage (Kim & Eustice, 2013). Algorithms generate 3D models of underwater archaeological sites using data from photogrammetry, LiDAR, and structure-from-motion techniques, allowing for virtual exploration and visualization.

The future entails harnessing increasingly complex convolutional neural networks (CNNs) trained on extensive datasets of underwater images. These advanced architectures have the potential to attain heightened performance in areas such as noise reduction, color correction, and object detection accuracy. These customized architectures can grasp the distinctive traits of underwater distortions, resulting in more resilient and efficient enhancement and detection algorithms. Generative Adversarial Networks (GANs) represent a category of deep learning models that could significantly transform underwater image enhancement (Guptha et al., 2022). By training GANs to produce lifelike, high-fidelity images from degraded underwater inputs, they have the capacity to effectively restore lost details and enhance image clarity. Expanding beyond the visible spectrum in underwater imaging extends to exploring infrared or ultraviolet wavelengths, unveiling concealed information. By integrating data from various spectral bands through sophisticated algorithms, there's the potential to gain a holistic understanding of the underwater environment, enhancing object detection and analysis capabilities. To enable autonomous underwater vehicles (AUVs) to make instantaneous decisions, rapid and efficient image processing algorithms, must be lightweight and efficient, are imperative. Researchers are investigating methods such as model compression and hardware acceleration to facilitate real-time image enhancement and object detection on AUV platforms with limited resources. With the growing complexity of algorithms, comprehending their decision-making process becomes increasingly significant. Explainable AI (XAI) methods are being researched to elucidate how image enhancement and object detection algorithms reach their conclusions, promoting trust and transparency in their usage (Dwivedi et al., 2023).

## 7.1 Ethical Considerations

Design of underwater image enhancement and object detection algorithm aids in the monitoring of exotic species, thereby preventing harm to ecosystem and native species. Enhanced images are crucial in monitoring the sensitive habitats of coral reefs and ensures the conservation and protection of such environment rather than exploiting them. Underwater object detection algorithms employed for surveillance purpose should be within the national security limits and must prevent the misuse of sensitive information. Another consideration is algorithmic bias, which can lead to inaccurate detection in case of search and rescue operations when trained on limited dataset. Military applications use underwater detection algorithms for submarine detection and mine detection. Non-combatant vessels must be detected and classified separately without leading to conflicts. Rather than developing separate algorithm for each application, same algorithm can be used for multiple purposes such as environmental monitoring can be reused for military purpose. The permanent deployment of underwater sensors and imaging devices such as sonar could disrupt marine ecosystem of species such as dolphin and whales. Such disruptions should be minimised for the welfare of marine organisms and their ecosystem.

## 8. CONCLUSION

The underwater realm's mesmerizing beauty and scientific significance are undeniable, yet capturing clear and informative images in this environment poses considerable challenges. Fortunately, computer-assisted algorithms have emerged as formidable tools, offering capabilities in image enhancement and object detection that overcome the constraints imposed by water's optical properties. These algorithms not only enhance the visual quality of underwater images but also empower researchers and explorers to delve deeper into analysing and understanding the underwater world. By tackling issues like color casts, low contrast, and limited light penetration, image enhancement algorithms enable accurate object detection. This integrated approach uncovers a wealth of information concealed within underwater images, supporting scientific studies, resource management initiatives, and underwater exploration. Challenges such as scarce training data, variations in real-world environments and computational complexities present exciting opportunities for further research and development. By fostering collaboration among researchers, data scientists, and industry experts, we can advance data collection methods, explore alternative learning approaches, and devise efficient and transparent algorithms. Sustained investment in research and development is essential to unlock the full potential of computer-assisted algorithms in underwater image processing. As we push the boundaries of this technology, we gain deeper insights into and appreciation for the underwater realm, paving the way for a more sustainable and knowledge-rich future for our oceans. The transformative power of innovative algorithms that illuminate the hidden marvels beneath the waves can be expanded with deep learning based approaches.

## REFERENCES


Buckland, M., & Gey, F. (1994). The relationship between recall and precision. *Journal of the American Society for Information Science*, *45*(1), 12–19. DOI: 10.1002/(SICI)1097-4571(199401)45:1<12::AID-ASI2>3.0.CO;2-L

Burns, C., Bollard, B., & Narayanan, A. (2022). Machine-learning for mapping and monitoring shallow coral reef habitats. *Remote Sensing (Basel)*, *14*(2666), 2666. Advance online publication. DOI: 10.3390/rs14112666

Chen, L., . . .. (2020). A benchmark dataset for both underwater image enhancement and underwater object detection. arXiv Preprint arXiv:2006.15789. https://doi.org//arXiv.2006.15789DOI: 10.48550

Chen, X., Zhang, Y., Li, D., & Duan, Q. (2023). Chinese mitten crab detection and gender classification method based on GMNet-YOLOv4. *Computers and Electronics in Agriculture*, *214*, 108318. Advance online publication. DOI: 10.1016/j.compag.2023.108318

Dai, J.. (2016). R-FCN: Object detection via region-based fully convolutional networks. *Proceedings of the 2016 Advances in Neural Information Processing Systems (NeurIPS)*, 379-387.

Dalal, N., & Triggs, B. (2005). Histograms of oriented gradients for human detection. *Proceedings of the 2005 IEEE Computer Society Conference on Computer Vision and Pattern Recognition (CVPR)*, 1, 886-893. DOI: 10.1109/CVPR.2005.177

Dosselmann, R., & Yang, X. D. (2006). Existing and emerging image quality metrics. *Proceedings of the 2006 Canadian Conference on Electrical and Computer Engineering (CCECE)*, 1906-1913. DOI: 10.1109/CCECE.2006.277467

Dwivedi, R., Dave, D., Naik, H., Singhal, S., Omer, R., Patel, P., Qian, B., Wen, Z., Shah, T., Morgan, G., & Ranjan, R. (2023). Explainable AI (XAI): Core ideas, techniques, and solutions. *ACM Computing Surveys*, *55*(9), 1–33. DOI: 10.1145/3561048

Er, M. J., Chen, J., Zhang, Y., & Gao, W. (2023). Research challenges, recent advances, and popular datasets in deep learning-based underwater marine object detection: A review. *Sensors (Basel)*, *23*(4), 1990. Advance online publication. DOI: 10.3390/s23041990 PMID: 36850584

Garcia, R., Nicosevici, T., & Cufí, X. (2002). On the way to solve lighting problems in underwater imaging. OCEANS'02 MTS/IEEE Conference, 2, 1010-1017. DOI: 10.1109/OCEANS.2002.1192107

Girshick, R. (2015). Fast R-CNN. *Proceedings of the 2015 IEEE International Conference on Computer Vision (ICCV)*, 1440-1448. DOI: 10.1109/ICCV.2015.169

Gomes, D., Saifuddin Saif, A. F. M., & Nandi, D. (2020). Robust underwater object detection with autonomous underwater vehicle: A comprehensive study. *Proceedings of the 2020 International Conference on Computing Advancements (ICCA)*, 1-8. DOI: 10.1145/3377049.3377052

Gonzalez-Rivero, M., Beijbom, O., Rodriguez-Ramirez, A., Bryant, D. E. P., Ganase, A., Gonzalez-Marrero, Y., Herrera-Reveles, A., Kennedy, E. V., Kim, C. J. S., Lopez-Marcano, S., Markey, K., Neal, B. P., Osborne, K., Reyes-Nivia, C., Sampayo, E. M., Stolberg, K., Taylor, A., Vercelloni, J., Wyatt, M., & Hoegh-Guldberg, O. (2020). Monitoring of coral reefs using artificial intelligence: A feasible and cost-effective approach. *Remote Sensing (Basel)*, *12*(3), 489. DOI: 10.3390/rs12030489

Gui, J., Cong, X., Cao, Y., Ren, W., Zhang, J., Zhang, J., Cao, J., & Tao, D. (2023). A comprehensive survey and taxonomy on single image dehazing based on deep learning. *ACM Computing Surveys*, *55*(13s), 1–37. DOI: 10.1145/3576918

Guptha, M. N.. (2022). Generative adversarial networks for unmanned aerial vehicle object detection with fusion technology. *Journal of Advanced Transportation*, *2022*, 5798423. Advance online publication. DOI: 10.1155/2022/7111248

Han, F., Yao, J., Zhu, H., & Wang, C. (2020). Underwater image processing and object detection based on deep CNN method. Hindawi. *Journal of Sensors*, *6740312*, 1–20. Advance online publication. DOI: 10.1155/2020/6707328

He, D. M., & Seet, G. G. L. (2002). Underwater lidar imaging in highly turbid waters. *Proceedings of SPIE - the International Society for Optical Engineering*, *4488*, 108–119. DOI: 10.1117/12.452805

He, K.. (2017). Mask R-CNN. *Proceedings of the 2017 IEEE International Conference on Computer Vision (ICCV)*, 2961-2969. DOI: 10.1109/ICCV.2017.322

Hou, X., Liu, J., Sun, Z., & Li, W. (2019). Underwater image enhancement via Retinex with illumination map transmission. *IEEE Access : Practical Innovations, Open Solutions*, *7*, 145207–145220. DOI: 10.1109/ACCESS.2019.2947384

Hu, S., & Liu, T. (2024). Underwater rescue target detection based on acoustic images. *Sensors (Basel)*, *24*(6), 1780. Advance online publication. DOI: 10.3390/s24061780 PMID: 38544042

Jaffe, J. S. (1990). Computer modeling and the design of optimal underwater imaging systems. *IEEE Journal of Oceanic Engineering*, *15*(2), 101–111. DOI: 10.1109/48.50695

Jerlov, N. G. (1976). *Marine optics*. Elsevier.

Kanimozhi, S., Gayathri, G., & Mala, T. (2019). Multiple real-time object identification using single shot multi-box detection. *Proceedings of the 2019 International Conference on Computational Intelligence in Data Science (ICCIDS)*, 1-6. DOI: 10.1109/ICCIDS.2019.8862041

Kaur, M., Kaur, J., & Kaur, J. (2011). Survey of contrast enhancement techniques based on histogram equalization. *International Journal of Advanced Computer Science and Applications*, *2*(7), 1–7. DOI: 10.14569/IJACSA.2011.020721

Kim, A., & Eustice, R. M. (2013). Real-time visual SLAM for autonomous underwater hull inspection using visual saliency. *IEEE Transactions on Robotics*, *29*(3), 719–733. DOI: 10.1109/TRO.2012.2235699

Kuzmin, K. A., & Grekov, N. A. (2021). Developing algorithmic software for automated monitoring of aquatic environment. *Proceedings of the 2021 IEEE Russian Automation Conference (RusAutoCon)*, 1-5. DOI: 10.1109/RusAutoCon52004.2021.9537506

Latif, G., Iskandar, D. A., Alghazo, J., Butt, M., & Khan, A. H. (2018). Deep CNN based MR image denoising for tumor segmentation using watershed transform. International Journal of Engineering & Technology, 7(2.3), 37-42.

Li, C., Anwar, S., Hou, J., Cong, R., Guo, C., & Han, J. (2020). An underwater image enhancement benchmark dataset and beyond. *IEEE Transactions on Image Processing*, *29*, 4376–4389. DOI: 10.1109/TIP.2019.2955241 PMID: 31796402

Li, Y., Liu, J., Li, X., & Tang, C. (2023). Underwater object detection algorithm based on adding channel and spatial fusion attention mechanism. *MDPI*, *11*(6), 822. DOI: 10.3390/mdpi11060822

Lin, T. Y., Maire, M., Belongie, S., Hays, J., Perona, P., Ramanan, D., Dollár, P., & Zitnick, C. L. (2014). Microsoft COCO: Common objects in context. *Proceedings of the 2014 European Conference on Computer Vision (ECCV)*, 740-755. DOI: 10.1007/978-3-319-10602-1_48

Liu, Z., Zhuang, Y., Jia, P., Wu, C., Xu, H., & Liu, Z. (2022). A novel underwater image enhancement algorithm and an improved underwater biological detection pipeline. *Journal of Marine Science and Engineering*, *10*(9), 1204. Advance online publication. DOI: 10.3390/jmse10091204

Lopez-Vazquez, V., Lopez-Guede, J. M., Chatzievangelou, D., & Aguzzi, J. (2023). Deep learning based deep-sea automatic image enhancement and animal species classification. *Journal of Big Data*, *10*(1), 37. Advance online publication. DOI: 10.1186/s40537-023-00711-w

Lu, H., & Serikawa, S. (2014). Underwater scene enhancement using weighted guided median filter. *Proceedings of the 2014 IEEE International Conference on Multimedia and Expo (ICME)*, 1-6. DOI: 10.1109/ICME.2014.6890157

Mäkynen, M., & Karvonen, J. (2017). Incidence angle dependence of first-year sea ice backscattering coefficient in Sentinel-1 SAR imagery over the Kara Sea. *IEEE Transactions on Geoscience and Remote Sensing*, *55*(11), 6355–6364. DOI: 10.1109/TGRS.2017.2721981

Medwin, H., & Clay, C. S. (1998). *Fundamentals of acoustical oceanography*. Academic Press.

Mobley, C., Stramski, D., Bissett, W., & Boss, E. (2004). Optical modeling of ocean waters: Is the Case 1–Case 2 classification still useful? *Oceanography (Washington, D.C.)*, *17*(2), 60–67. DOI: 10.5670/oceanog.2004.48

Mobley, C. D. (1994). *Light and water: Radiative transfer in natural waters*. Academic Press.

Ngo, D., Lee, G. D., & Kang, B. (2019). Improved color attenuation prior for single-image haze removal. *Applied Sciences (Basel, Switzerland)*, *9*(19), 4011. Advance online publication. DOI: 10.3390/app9194011

Parthasarathy, S., & Sankaran, P. (2012). An automated multi scale retinex with color restoration for image enhancement. *Proceedings of the 2012 National Conference on Communications (NCC)*, 1-5. DOI: 10.1109/NCC.2012.6176791

Patel, Z., Desai, C., Tabib, R. A., Bhat, M., Patil, U., & Mudengudi, U. (2020). Framework for underwater image enhancement. *Procedia Computer Science*, *171*, 491–497. DOI: 10.1016/j.procs.2020.04.052

Peng, Y. T., Cao, K., & Cosman, P. C. (2018). Generalization of the dark channel prior for single image restoration. *IEEE Transactions on Image Processing*, *27*(6), 2856–2868. DOI: 10.1109/TIP.2018.2813092 PMID: 29570087

Pizer, S. M., Amburn, E. P., Austin, J. D., Cromartie, R., Geselowitz, A., Greer, T., ter Haar Romeny, B., Zimmerman, J. B., & Zuiderveld, K. (1987). Adaptive histogram equalization and its variations. *Computer Vision Graphics and Image Processing*, *39*(3), 355–368. DOI: 10.1016/S0734-189X(87)80186-X

Priyadharsini, R., Sree Sharmila, T., & Rajendran, V. (2018). A wavelet transform based contrast enhancement method for underwater acoustic images. *Multidimensional Systems and Signal Processing*, *29*(5), 1845–1859. DOI: 10.1007/s11045-017-0533-5

Ren, Z., Zhang, Q., Gao, X., Hao, P., & Cheng, J. (2021). Multi-modality learning for human action recognition. *Multimedia Tools and Applications*, *80*(11), 16185–16203. DOI: 10.1007/s11042-019-08576-z

Shamrat, J. M., . . .. (2022). Human face recognition applying Haar cascade classifier. In M. Tiwari & S. K. Gupta (Eds.), Pervasive Computing and Social Networking: Proceedings of ICPCSN 2021 (pp. 401-411). Springer. https://doi.org/DOI: 10.1007/978-981-16-4440-1_39

Sung, M., Yu, S. C., & Girdhar, Y. (2017). Vision based real-time fish detection using convolutional neural network. *Proceedings of the 2017 OCEANS Conference (Aberdeen)*, 1-5. DOI: 10.1109/OCEANSE.2017.8084889

Tarekegn, A., Alaya Cheikh, F., Ullah, M., Sollesnes, E., Alexandru, C., Azar, S., Erol, E., & Suciu, G. (2023). Underwater object detection using image enhancement and deep learning models. *Proceedings of the 2023 European Workshop on Visual Information Processing (EUVIP)*, 1-6. DOI: 10.1109/EUVIP58404.2023.10323047

Terven, J., & Cordova-Esparza, D. (2023). A comprehensive review of YOLO architectures in computer vision: From YOLOv1 to YOLOv8 and YOLO-NAS. arXiv Preprint arXiv:2304.00501. https://doi.org//arXiv.2304.00501DOI: 10.48550

Tritech International Limited. (2008). *Side scan sonar*. Tritech International Limited.

Wang, X., Xue, G., Huang, S., & Liu, Y. (2023). Underwater object detection algorithm based on adding channel and spatial fusion attention mechanism. *Journal of Marine Science and Engineering*, *11*(1116), 1116. Advance online publication. DOI: 10.3390/jmse11061116

Wang, Z., Zhu, J., Yan, T., & Yang, L. (2019). A new modified wavelet-based ECG denoising. Computer Assisted Surgery, 24(sup1), 174-183. DOI: 10.1080/24699322.2018.1560088

Wei, J., Wang, S., & Huang, Q. (2020). F$^3$Net: Fusion, feedback and focus for salient object detection. *Proceedings of the 2020 AAAI Conference on Artificial Intelligence (AAAI)*, 34(7), 12315-12322. DOI: 10.1609/aaai.v34i07.6916

Woon, S. G. (2020). Digital image enhancement. In *Digital Image Processing*. Springer., DOI: 10.1007/978-981-10-5550-8_12

Yang, Y., Liang, W., Zhou, D., Zhang, Y., & Xu, G. (2023). Object detection for underwater cultural artifacts based on deep aggregation network with deformation convolution. *Journal of Marine Science and Engineering*, *11*(2228), 2228. Advance online publication. DOI: 10.3390/jmse11122228

Zhang, D., He, Z., Zhang, X., Wang, Z., Ge, W., Shi, T., & Lin, Y. (2023). Underwater image enhancement via multi-scale fusion and adaptive color-gamma correction in low-light conditions. *Engineering Applications of Artificial Intelligence*, *126*, 106972. Advance online publication. DOI: 10.1016/j.engappai.2023.106972

Zhang, W., Li, G., & Ying, Z. (2017). A new underwater image enhancing method via color correction and illumination adjustment. *Proceedings of the 2017 IEEE Visual Communications and Image Processing (VCIP)*, 1-4. DOI: 10.1109/VCIP.2017.8305027

Zuiderveld, K. (1994). Contrast limited adaptive histogram equalization. In Heckbert, P. S. (Ed.), *Graphics Gems IV* (pp. 474–485). Academic Press., DOI: 10.1016/B978-0-12-336156-1.50061-6

# Chapter 5
# Deep Learning Strategies for Analyzing EEG Signals in Emotion Recognition

**K. V. Kavitha**
*Department of Computer Science and Engineering, Annamalai University, India*

**L. R. Sudha**
*Department of Computer Science and Engineering, Annamalai University, India*

**J. S. Jayasudha**
*Department of Computer Science, Central University, India*

## ABSTRACT

*Emotions are vital in human communication, and EEG signals have become a potential method for detecting emotional states in real-time. Nevertheless, attaining a high level of precision in the identification of emotions continues to be a significant obstacle. Due to the emergence of deep learning techniques, there has been a growing desire to utilize these approaches for acquiring advanced characteristics in EEG emotion identification. This research makes a valuable contribution to the field of Human-Computer Interaction (HCI) medical applications. The purpose of this chapter is to offer a thorough examination of EEG emotion identification, specifically emphasizing the use of deep learning techniques. The initial focus of our study is a comprehensive exploration of the underlying principles and ideas presented in the existing literature. The primary objective of our research is to enhance the efficiency of emotion recognition models to provide more precise outcomes. This chapter offers insights into the most recent breakthroughs and potential areas for further inquiry.*

DOI: 10.4018/979-8-3693-3840-7.ch005

## INTRODUCTION

Emotions constitute a fundamental aspect of human experience, influencing our feelings, behaviors, and communications. Emotion recognition points to the ability to identify and infer human emotions based on various such as countenance, vocal accent, physiological signals (like heart rate or skin conductance), and behavioral patterns. It involves understanding and categorizing emotions, often into basic categories like happiness, sadness, anger, fear, and disgust, but can also include more nuanced emotions or complex emotional states. Emotion recognition using EEG would involve using patterns of brain activity to identify and categorize specific emotional states. EEG offers a promising avenue to explore emotions by capturing brain activity, providing insights into the neural correlates underlying different emotional responses (Alarcão & Fonseca, 2019).

The field of emotion recognition has garnered significant attention in recent years, mostly because of its potential applications in healthcare, neuromarketing, and human-computer interaction. Electroencephalogram (EEG) is a highly promising technology for evaluating emotional states since it has a high temporal resolution and is directly linked to brain processes. This research investigates the utilization of deep learning techniques to analyze EEG signals with the aim of reliably and effectively identifying human emotions (Erat et al., 2024).

Deep learning, a subfield of machine learning, has demonstrated exceptional achievements in representing intricate patterns in data across various fields. Deep learning approaches in the context of EEG recordings can uncover complex patterns that may be challenging to identify using conventional signal processing methods. This paper explores different deep learning architectures, including Convolutional Neural Networks (CNNs), Recurrent Neural Networks (RNNs), and hybrid models that contain elements of both. It offers a detailed analysis of their abilities and constraints in the field of emotion recognition.

The examination commences by examining the essential attributes of EEG signals that are pertinent to emotional processing. This is then followed by a discourse on the necessary pre-processing measures to ready EEG data for deep learning applications. Following that, the research assesses various deep learning models that have been suggested for emotion recognition based on EEG data, emphasizing their performance measures, such as accuracy, precision, and recall, in different experimental configurations. Furthermore, this analysis thoroughly investigates the difficulties related to the implementation of these models, including handling data with many dimensions, requiring significant processing resources, and assuring that the models can be applied to diverse persons.

This paper aims to provide a comprehensive analysis of deep learning strategies for EEG-based emotion recognition. It seeks to demonstrate how these advanced computational techniques can be utilized to improve our comprehension of human emotions, leading to the development of innovative applications that can have various societal benefits.

## EEG

EEG, or electroencephalography, is a diagnostic technique used to record the impulse of the brain. It is done by placing electrodes on the scalp to measure the impulse produced by brain cells, or neurons. These signals, called brainwaves, reflect the functioning of the brain, and can provide intuition into various cognitive processes, including emotions **(LIU ET AL., 2016)**.

Emotion detection using EEG relies on the fact that different emotional states are associated with distinct patterns of brain activity. By analyzing these patterns, researchers can infer the emotional state of an individual. For example, heightened activity in certain brain regions may indicate feelings of happiness or excitement, while decreased activity may signal sadness or anxiety. EEG offers several advantages for emotion detection, including real-time monitoring, high temporal resolution, and low cost to a certain extent compared to other neuroimaging techniques like fMRI. However, it also has constraints, such as lower spatial resolution and prone to artifacts from muscle activity or environmental noise **(BABY, 2014;CRAIK ET AL., 2019)**.

In terms of safety, EEG is considered non-invasive and generally harmless. The electrodes used in EEG are placed on the scalp's surface and do not involve penetration into the brain or any exposure to radiation. Therefore, EEG is widely used in both research and clinical settings for studying emotions, cognitive processes, and neurological disorders. However, as with any medical procedure, proper precautions should be taken to ensure the safety and comfort of the individual undergoing EEG recording.

EEG recording is done by placing electrode on the scalp to detect the electrical signals. These signals are normally weak. So, they are amplified. Filtering is done to remove unwanted noise. These signals are recorded by computer and displayed as brain waves.The brain waves associated with EEG recordings are Delta, Theta, Alpha,Beta and Gamma waves.These EEG recordings are analysed to identify abnormal patterns in brain or diagnose neurological disorders.The slowest brain wave is the delta wave whose frequency is in the range 0.5 to 4 Hz. They are associated with restorative sleep and unconscious bodily functions. Theta waves(4-8Hz) are associated with emotional processing and accessing the subconscious mind. During deep relaxation these waves are active. Alpha waves(8-12Hz) have low amplitude. During relaxed state of brain, the Beta waves (12-30 Hz) are prominent and characterized by higher frequencies and amplitudes. Beta waves are associated with abstract thought, concentration, and active brainstorming. The fastest brain wave is the Gamma wave(30-100Hz). During tasks which require deeper thinking, awareness and memory formation these waves are noticed. The patterns associated with brain waves can provide insights into mental condition and neurological disorders.

The basic steps in emotion recognition using EEG are data collection, extracting features to analyze the amplitude, frequency or synchronization of brain waves in various regions of the brain, applying machine learning or deep learning algorithms to connect brain patterns with specific emotional states and the last step is validation using independent datasets and refinement.

## Purpose

The purpose of conducting emotion analysis is to delve into the intricate landscape of human emotions, seeking to objectively measure, interpret, and understand the nuanced spectrum of emotional states. By employing various tools and methodologies, such as neuroimaging techniques like EEG, the aim is to decode the underlying neural patterns associated with different emotions. This pursuit serves to provide valuable insights into human behavior, decision-making processes, and social interactions. It outlines the system's purpose in objectively assessing emotional states, describes the interfaces utilized for data collection and analysis, elucidates the constraints within which the system operates, and delineates how the system interprets neural stimuli to recognize and categorize diverse emotional responses in real-time(Lakshminarayanan et al., 2023).

## Scope of the System

Emotion classification focuses on four distinct emotional states commonly encountered in human experiences: neutral, sad, fear, and happy. These emotions represent a diverse range of affective states and are fundamental to understanding human behavior and cognition domains. By leveraging EEG data, meaningful features are extracted that can be used to train deep learning models for emotion classification.

To achieve the classification task, emotion detection explores the capability of various deep learning architectures, including Conv3D, LSTM, and CONVLSTM2D. These models offer different approaches to capturing temporal dependencies and spatial patterns in EEG data, allowing for a comprehensive exploration of emotion classification techniques. Training is conducted on publicly available EEG datasets, ensuring reproducibility and comparability of results. The models are assessed using standard classification metrics such as precision, recall, F1-score, and accuracy.

Human brain activity is reflected by emotion, and the four basic emotions are happy, sad, fear an d surprise. Classification of emotions is based on time and frequency domain EEG features. The commonly used techniques for emotion recognition include Convolutional Neural Network (CNN), Recurrent Neural Network (RNN), Deep Belief Network, Long Short-term Memory Neural Network (LSTM) and BiLSTM.

## CONVOLUTION NEURAL NETWORK

CNN **(Albawi et al., 2017)**works by applying convolution operations to input data using filters. Filters slide over the input data and capture spatial patterns like corners, edges and shapes. After the convolution operation activation function ReLU is applied to initiate non-linearity and learn complex patterns. Pooling layers are used to under sample the feature maps and lower computational complexity and to prevent overfitting. Fully connected layers are used to perform classification. Softmax function is normally used in this layer to generate class probabilities for classification tasks **(Ravikumar et al., 2022;Ravikumar & Sriraman, 2024b)**.

*Figure 1. Convolution neural network*

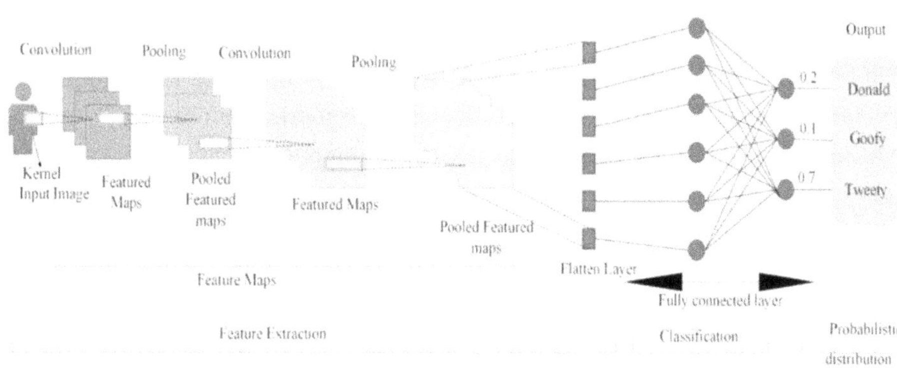


## Convolutional-3D Model

The Conv3D model is a variant of convolutional neural networks (CNNs) intended to handle three-dimensional data, making it appropriate for processing spatio- temporal information.

*Table 1. Summary of the Conv3D layers*

| Layer (type) | Output Shape | Param # |
|---|---|---|
| conv3d (Conv3D) | (None, 64, 8, 9, 32) | 80,032 |
| batch normalization | (None, 64, 8, 9, 32) | 128 |
| activation (Activation) | (None, 64, 8, 9, 32) | 0 |
| dropout (Dropout) | (None, 64, 8, 9, 32) | 0 |
| max pooling3d (MaxPooling3D) | (None, 32, 4, 4, 32) | 0 |
| flatten (Flatten) | (None, 16384) | 0 |
| dense (Dense) | (None, 256) | 4,194,560 |
| batch normalization 1 | (None, 256) | 1,024 |
| activation 1 (Activation) | (None, 256) | 0 |
| dropout 1 (Dropout) | (None, 256) | 0 |
| dense 1 (Dense) | (None, 10) | 2,570 |

The model is structured as a sequential neural network, incorporating layers designed to extract features from EEG signals and perform emotion detection. Starting with a Conv3D layer, which applies 3D convolutional operations to retrieve spatial and temporal dependencies within the EEG signal, the architecture includes subsequent layers such as batch normalization, activation, dropout, max-pooling3D, and dense layers. These layers collectively contribute to feature extraction and emotion classification. With a total of 4,278,314 parameters, the model consists of both trainable and non-trainable parameters, enabling it to optimize performance for EEG signal analysis. For EEG signals, the Conv3D model operates by leveraging convolutional operations to capture spatial and temporal dependencies, extracting discriminative features indicative of different emotional states, and facilitating accurate emotion classification.

*Figure 2. Convolutional-3D model*

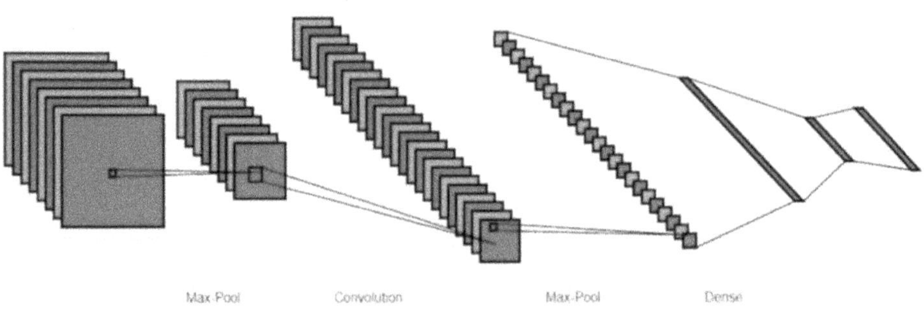

## LSTM and Bi-LSTM Model

The Long Short-Term Memory (LSTM) (Hochreiter & Schmidhuber, 1997)model belongs to the family of recurrent neural networks (RNNs) designed to capture sequential dependencies in data. LSTMs are particularly effective in modeling time series data due to their ability to retain long-term dependencies while mitigating the vanishing gradient problem. Structured as a sequential neural network, the model comprises bidirectional LSTM layers, dropout layers for regularization, dense layers, and batch normalization for stabilizing training. With a total of 366,372 parameters, the model is adept at capturing temporal dependencies in EEG signals, facilitating accurate emotion classification.

*Table 2. Summary of the LSTM layers*

| Layer (type) | Output Shape | Param # |
|---|---|---|
| bidirectional (Bidirectional) | (None, 1440, 256) | 197,632 |
| dropout (Dropout) | (None, 1440, 256) | 0 |
| bidirectional 1 (Bidirectional) | (None, 128) | 164,352 |
| dropout 1 (Dropout) | (None, 128) | 0 |
| dense (Dense) | (None, 32) | 4,128 |
| batch normalization (BatchNormalization) | (None, 32) | 128 |
| dense 1 (Dense) | (None, 4) | 132 |

For EEG signal analysis, the bidirectional LSTM layers are pivotal components, enabling the model to effectively capture temporal dynamics and long-term dependencies present in the EEG data. By processing the input data bidirectionally, the model can capture both past and future context, enhancing its ability to extract relevant features indicative of different emotional states.

The dropout layers introduced after each LSTM layer aid in preventing overfitting by erratically dropping a fraction of the input units during training, thereby improving the generalization performance of the model. Additionally, batch normalization is applied to balance and facilitate the training process by normalizing the activations of the previous layer. Following the LSTM layers, dense layers are utilized for feature extraction and emotion classification. These layers transform the extracted features into probabilities for various emotional categories. Overall, the LSTM model operates by leveraging its bidirectional nature to capture temporal dependencies in EEG signals, ultimately enabling accurate and robust emotion classification.

*Figure 3. LSTM*

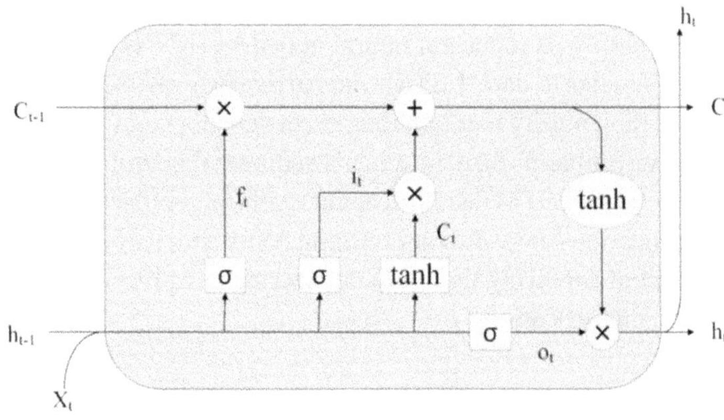


## Convolutional-LSTM2D Model

The ConvLSTM2D model combines the spatial convolutional potentiality of CNNs with the sequential modeling capabilities of LSTMs, making it well-suited for processing spatiotemporal data.

*Table 3. Summary of the ConvLSTM2D layers*

| Layer (type) | Output Shape | Param # |
|---|---|---|
| conv lstm2d 7 (ConvLSTM2D) | (None, 8, 9, 16) | 33,664 |
| dropout 1 (Dropout) | (None, 8, 9, 16) | 0 |
| flatten 2 (Flatten) | (None, 1152) | 0 |
| dense 4 (Dense) | (None, 256) | 295,168 |
| dense 5 (Dense) | (None, 10) | 2,570 |

Structured as a sequential neural network, the model consists of a ConvLSTM2D layer, a dropout layer for regularization, a flatten layer to modify the output, and two dense layers for feature extraction and emotion classification. With a total of 331,402 parameters, the model effectively encapsulates both spatial and temporal dependencies in EEG data, facilitating accurate emotion analysis.

For EEG signal analysis, the ConvLSTM2D layer serves as a critical component, combining the spatial convolutional capabilities of CNNs with the sequential modeling capabilities of LSTMs. This unique architecture enables the model to encapsulate both spatial and temporal patterns in EEG signals simultaneously. By processing EEG data through the ConvLSTM2D layer, the model can extract intricate spatial-temporal features indicative of different emotional states.

The dropout layer introduced after the ConvLSTM2D layer aids in preventing overfitting by randomly dropping a snippet of the input units during training. This regularization technique enhances the generalization performance of the model and reduces the risk of memorizing noise in the training data.

Following the dropout layer, the flatten layer modifies the output from the convLSTM2D layer into a one-dimensional array, arranging it for input into the subsequent dense layers. These dense layers are responsible for feature extraction and emotion classification, transforming the extracted features into probabilities for various emotional categories.

Overall, the ConvLSTM2D model operates by leveraging its unique architecture to capture both spatial and temporal dependencies in EEG signals, enabling accurate and robust analysis of emotion. By extracting informative spatial-temporal features, the model contributes to the comprehensive understanding of brain activity associated with different emotional states, ultimately enhancing the accuracy of emotion classification in EEG signals (Ravikumar & Sriraman, 2023a, 2023b, 2024a).

## Recurrent Neural Network

It is a type of neural network where the previous stage output is fed as input to current stage. RNN has a hidden layer which remembers some information about a sequence(Gelenbe, 1993).

*Figure 4. RNN*

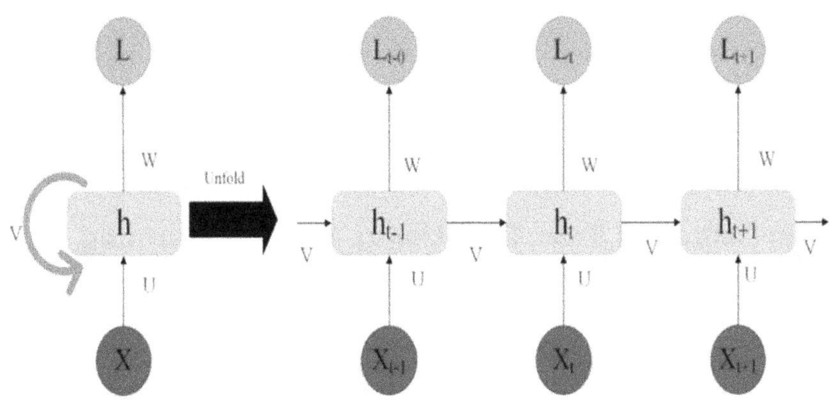

## Deep Belief Network

Deep belief networks consist of various layers of neurons, each connected to the neuron of the subsequent layer. Here the layers are connected to one another, and there are no intra-layer connections. Each layer trains independently on the output of the previous one. So, it can be considered as a separate model. A DBN is a stack of networks, each of which has learned different features and traits from the raw data. DBNs can work on both supervised and unsupervised problems due to complex architecture. Moreover, their ability to comprehend complex underlying data patterns makes them excellent for generative applications.

*Figure 5. Deep belief network*

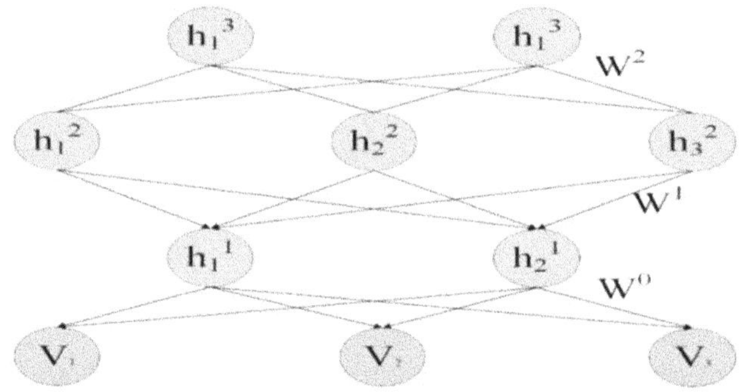


## Emotion Detection From EEG Signals

In recent years, numerous systems for emotion analysis using EEG have emerged across both developed and developing countries, showcasing a growing interest and investment in leveraging neuroimaging technology to decipher and interpret human emotions.

## Emotion Detection From EEG Using Transfer Learning

The landscape of EEG-based emotion classification has seen notable advancements, particularly in response to challenges related to limited data availability. Traditional methods, such as Short Time Fourier Transform (STFT), Power Spectral Density (PSD), and Continuous Wavelet Transform (CWT), have been pivotal but faced constraints in achieving optimal performance. Recent studies have emphasized the importance of feature combination for a more comprehensive representation of brain activity during emotional states. In this context, innovative features like Mean Phase Coherence (MPC), Magnitude Squared Coherence (MSC), and Differential Entropy (DE), aiming to capture different facets of EEG data are strategically combined to create a robust representation of emotional states.

Transfer learning emerges as a key paradigm in EEG-based emotion classification, leveraging knowledge from one domain to enhance performance in another. The application of transfer learning is exemplified in the use of a pre-trained ResNet-50 model, originally designed for image classification on the ImageNet dataset. This model, with its hierarchical feature learning capabilities, has demonstrated efficacy in extracting relevant features from EEG data, contributing to improved accuracy and efficiency in emotion classification.

It reported high accuracies in subject-dependent and subject-independent emotion classification underscore the potential of the proposed methodology. By surpassing chance accuracy and outperforming traditional methods, the study suggests the effectiveness of the combined features and transfer learning approach. These findings not only advance the understanding of EEG-based emotion classification but also hold promise for practical applications, including neuroproteins and mental health interventions.

Looking ahead, it suggests the future research directions, including the exploration of data augmentation techniques, enhanced classifiers, and additional features to further improve the accuracy and generalization of EEG-based emotion classification models. Transfer learning approach contributes significantly to the field by introducing an innovative approach that amalgamates advanced features and transfer learning for enhanced EEG-based emotion classification, paving the way for broader applications and future advancements.

## Multi-Class Emotion Classification Using EEG Signals

Emotion detection using EEG signals and deep learning has seen significant growth in recent years. Several studies have leveraged advanced machine learning techniques to decode emotional states based on brain activity. The utilization of deep learning models, including Long Short-Term Memory (LSTM) and Convolutional Neural Network (CNN), aligns with a broader trend in exploring the capabilities of neural networks for EEG analysis. Previous research has demonstrated the potential of EEG signals in capturing neural correlates of emotions. These studies often emphasize the importance of feature extraction methods to reveal relevant patterns in the EEG data. While deep learning models are known for their capability to abruptly learn hierarchical representations, the incorporation of Fast Fourier Transform (FFT) for feature extraction,introduces an interesting fusion of traditional signal processing techniques with modern neural networks.

The choice of the DEAP dataset reflects a common trend in utilizing publicly available datasets for benchmarking emotion recognition algorithms. However, it recognizes the need for diverse datasets and real-world scenarios to enhance the generalizability of emotion classification models. Addressing this limitation is crucial for ensuring the practical applicability of such models across various contexts.

Moreover, the comparative analysis between LSTM and CNN models is reviewed which echoes a broader interest in assessing the strengths and weak- nesses of different deep learning architectures for EEG-based emotion classification. This comparison contributes valuable insights into the suitability of these models for specific tasks and informs future research directions.

The existing body of research consistently underscores concerns regarding the interpretability of deep learning models. Scholars are increasingly acknowledging the imperative for transparent models, especially in critical domains such as healthcare. Despite the demonstrated high accuracy, it is evident that broader challenges within the field persist. Notably, the absence of comprehensive discussions on model interpretability and hyperparameter tuning raises questions about the generalizability of the findings and suggests the need for a more nuanced exploration of these aspects in future research.

EEG-based emotion classification with deep learning demonstrates a dynamic interplay between traditional signal processing techniques and modern neural networks. Addressing challenges related to dataset diversity, model interpretability, and hyperparameter optimization remains essential for advancing the field and fostering practical applications in areas such as mental health diagnosis and human computer interaction.

## Consumer Emotional State Evaluation Using Deep Learning Approach for EEG Based Emotion Recognition

Major steps involved in the process of using EEG for emotion recognition are signal acquisition, pre-processing (removing artifacts), feature extraction using Power Spectral Density (PSD), and classification using SVM, k-NN, RF, and DNN algorithms. Classifiers are compared for emotional state detection using EEG data from the DEAP dataset. The EEG data from the DEAP database were pre-processed by down sampling to 128 Hz, applying bandpass filters (4.0 Hz to45.0 Hz), and eliminating EOG artifacts using Independent Component Analysis (ICA). ICA helped extract useful features and remove noise, retaining channels (Fz, F3, F4, AF3, and AF4). Feature extraction from EEG signals is crucial for emotion recognition. Power Spectral Density (PSD), a frequency domain technique is used to fragment

EEG frequency bands (alpha, beta, gamma, theta). PSD helps identify emotional states, and the extracted frequency bands were used to calculate valence. The Deep Neural Network (DNN) architecture comprise of input, three hidden layers, batch normalization, and an output layer. The DNN was trained using normalized features and the Adam gradient descent optimization strategy. Various loss functions like cross entropy were applied, and the system was stopped upon overfitting. The DNN model's architecture, layer sizes, activation functions (ReLU), and output (Softmax) were detailed along with the usage of min-max normalization for data processing.

The utilization of EEG data is highlighted to extract PSD and valence signals, resulting in around 2125 features per EEG activity. The performance of four classifiers (DNN, SVM, KNN, RF) are evaluated using various validation techniques (LOOCV, holdout, K-fold) and reports their accuracy scores. The proposed method achieves high accuracy rates of around 70%, 93%, 91%, 84%, and 87% across different validation parameters, outperforming other methods. Focuses on only two emotional states and employing a limited number of evaluation parameters. Additionally, the DNN model's applicability can be extended to multiple modalities to better understand consumer emotional states.

## EEG-Based Emotion Detection With Deep Convolution Neural Network

The DEAP dataset contains self-assessment ratings and experimental data, featuring EEG and physiological recordings, participant ratings, questionnaires, face videos for 22 participants, and raw physiological data. Preprocessed to Matlab/Python formats, each participant's 40 experiments provide 8064 EEG/physiological signals from 40 channels.

Utilizing dimensionality reduction via statistical features, 101 values per channel are extracted from each experiment. To manage the large data, each experiment's 8064 readings are divided into ten batches, extracting 90 processed values for each batch. These are combined with net statistics and experiment details to yield 101 values per channel. A statistical reduction method shortens the dataset from 322,560 readings to 4040 values per experiment, forming the initial processed dataset for training and cross-validation.

Two neural models, a Deep Neural Network (DNN) and a Convolutional Neural Network (CNN), are employed for the research, incorporating modern machine learning techniques like Dropout and Rectilinear Units. The DNN comprises four fully connected layers, starting with an input layer taking 4040 readings per experiment. Successive layers reduce the dimensionality to yield 2 or 3 classes (one-hot encodings) for classification. ReLU activation functions are applied, alongside Dropout techniques, to ensure non-linearity and model accuracy. The learning process involves

RMSProp for gradient descent, using a learning rate of 0.00001. ReLU activation is chosen for its biological plausibility, and the dropout probabilities (0.25 and 0.5) are empirically determined for optimal results. This model employs cross-entropy to minimize differences between actual and predicted outcomes, aiming for better learning from limited datasets.

CNNs, efficient for image classification, are adapted for DEAP data representation. Each experiment's data is represented as a 2D array (40 * 101) for the 40 channels with 101 readings each. The CNN architecture starts with convolution layers followed by MaxPooling layers, Dropout techniques, and fully connected dense layers. Activation functions like Tan Hyperbolic and Rectilinear units are crucial, affecting neuron activation for later layers. Each layer's configuration involves filter counts, kernel sizes, and activation functions specifically optimized for valence and arousal classification. The CNN model utilizes categorical cross-entropy for loss calculation and stochastic gradient descent (SGD) as the optimizer, adjusting learning rates (0.00001 for Valence,0.001 for Arousal) to improve model training. The training is conducted in batches with epochs set at 250.The results section compared the two models for emotion classification accuracy. The DNN model achieved around 75-82percent accuracy for Valence and Arousal classifications in 2 classes. On the other hand, the CNN model showed higher accuracy, around 81-73percent, demonstrating better performance in emotion classification. For 3-class classification, both models outperformed previous methods. The CNN model generally had more variable accuracies across subjects compared to the DNN model, which displayed more consistent results. Overall, these models showcased significant progress in using neural networks for EEG-based emotion recognition.

## Emotion Recognition of EEG Signals With Machine Learning Approach

Here a robust emotion recognition system is developed utilizing Electroencephalography (EEG) signals based on the valence-arousal model. EEG signals, known for their non-linear, non-stationary nature and susceptibility to various noise sources, were the focus of investigation. The study utilized the

DEAP database, containing EEG data from 32 participants exposed to music videos and labeled for valence, arousal, dominance, and liking. Channels associated with frontal brain regions, specifically F3-F4, F7-F8, FC1-FC2, FC5- FC6, and FP1-FP2, were chosen for analysis, considering their link to positive and negative emotions.

To ensure signal quality, the Average Mean Reference (AMR) method was applied for noise reduction, and normalization between 0 and 1 was conducted to mitigate individual differences. The EEG signals were segmented into 2- and 4-second intervals with a 50-percentage overlap, allowing for rapid detection of emotional states. The Discrete Wavelet Transform (DWT) was employed to decompose signals into five frequency bands: theta, alpha, beta, gamma, and noise. From these bands, spectral features, entropy, and energy were extracted from each window within the frequency bands, aiming to capture diverse aspects of the EEG data that could signify emotional states.

Following feature extraction, Principal Component Analysis (PCA) was employed. PCA, a technique for dimensionality reduction, transformed the extracted features into uncorrelated attributes, reducing dimensionality while retaining relevant information. Three classification algorithms – Support Vector Machine (SVM), K-Nearest Neighbor (KNN), and Artificial Neural Network (ANN) – were utilized for emotional state classification. The classifiers were trained with features from different frequency bands and channel pairs, employing eight-fold cross-validation to ensure robustness.

The research yielded promising results, showcasing notable accuracy levels in emotion classification. The cross-validated SVM with a radial basis function (RBF) kernel achieved an accuracy of 91.3 per for arousal and 91.1 per for valence, specifically in the beta frequency band. Compared to existing algorithms applied to the DEAP dataset, the proposed approach demonstrated improved performance in emotion classification based on the valence-arousal model. Overall, the study contributed insights into effectively utilizing advanced signal processing techniques, such as DWT and PCA, in conjunction with machine learning algorithms for accurate emotion recognition from EEG data.

The findings hold significance for cognitive science and brain-computer interface systems, paving the way for enhanced understanding and utilization of emotional signals in various applications.

## Emotion Detection of EEG Signals Using CNN on DEAP Dataset

Emotions, being intricate and multifaceted, lack clear-cut boundaries, affected by environmental, linguistic, cultural, and situational nuances. This complexity renders the recognition of human emotions exceedingly challenging. Each subject watched 15 Chinese film clips designed to elicit positive, neutral, and negative emotions across three sessions, resulting in a total of 45 experiments. The EEG signals and facial videos of the subjects were recorded while they viewed the emotional film clips, sampled at 200Hz with an EEG cap containing 62 channels based on the international 10-20 system.

The process of converting EEG signals into image data for CNN processing is cumbersome and seemingly adapted to suit the algorithm, prompting a different approach. The primary focus revolves around bypassing the intermediary step of signal-to-image transformation. Instead, direct utilization of windowed EEG signals as input for the CNN, aiming to enable the network to extract pertinent information directly from the EEG data. To facilitate CNN training with limited data, the EEG signals were segmented into smaller time segments to enhance training and real-time performance. However, due to the short duration of these segments, direct feature extraction was not performed, and 50x62 data graphs were directly input into the CNN.

In a Convolutional Neural Network (CNN), the crucial component is the convolution kernel, typically optimized for image data but challenging to adapt to the unique nature of EEG signals. Unlike images, EEG data is time-based and strongly correlated with temporal sequences. To address this, specialized kernels tailored for EEG, focusing on time, signal channels, and a combination of both are used. However, due to the limited correlation between time and space in EEG signals, the combined kernel was deemed unsuitable. Instead prioritizes temporal information before spatial aspects within the CNN architecture. By independently processing time and space in different network layers, the model becomes more adaptable, effectively navigating the intricacies of EEG data analysis. CNN uses three different convolution kernels. The first convolution kernel is only designed for the time domain and the kernel shape is 1*m while the 'm' represents the time related convolution length. The second convolution kernel is only designed for the signal channels and the size is n*1 while n represents the channels relate convolution length. The third kernel size is n*m and this kind of kernel is the mix of channels convolution and time domain convolution. The CNN structure involves a network with 7 layers, including two convolutional layers and two fully connected layers with 512 neurons each. The network's output layer comprises three neurons for emotion classification.

The results show varied accuracies among subjects, hence adjustments in convolutional kernels may be necessary per subject. Additionally, SVM and DNN classifiers were employed with different feature extractions from the same dataset, achieving an accuracy of 0.8608. Not all 62 EEG channels are pertinent to emotional control, four electrode profiles are designed, targeting brain areas involved in emotional regulation. However, CNN's modeling capability experiences a significant drop when channels are reduced. Specifically, while the 12-channel profile performs well with SVM, CNN's accuracy notably decreases, indicating CNN's need for extensive EEG data and its sensitivity to noise reduction.

## Emotion Detection from EEG Recordings

This study delves into the complex realm of emotion recognition, specifically focusing on the intricate relationship between brain signals, physiological responses, and human emotional states. It centers on leveraging a diverse dataset, the DEAP

(Database for Emotional Analysis using Physiological Signals), to unravel the correlation between EEG signals and emotions. The study employs a multifaceted approach, extracting various features from EEG recordings using different domains—time, frequency, time-frequency, and multi- electrode features—to comprehensively capture and represent the underlying physiological aspects linked to emotions.

The methodology encompasses a meticulous feature selection process employing the mRMR technique. This approach strategically identifies pertinent features that exhibit maximum relevance to emotions while minimizing redundancy, enhancing the accuracy and efficiency of subsequent classification models. These chosen features act as crucial indicators for discerning distinct emotional states within the dataset, paving the way for a more nuanced understanding of the interplay between brain activities and emotions.

This study employs a multifaceted approach, harnessing machine learning techniques, feature extraction from EEG signals, and clustering methodologies like K-means. Through feature extraction methods encompassing diverse domains such as time, frequency, time-frequency, and multi-electrode features, the study captures intricate nuances within EEG recordings. These features, spanning statistical calculations, oscillatory patterns, signal shapes, and more, are meticulously selected using the Maximum Relevance Minimum Redundancy (mRMR) method, enhancing the model's predictive power. Additionally, K- means clustering aids in revealing underlying structures within the dataset, illuminating distinctive patterns characterizing various emotional states. This clustering process offers a foundation for deeper insights into the complex interplay between brain signals and emotions, providing a refined understanding of the emotional landscape captured in the DEAP dataset.

The research then employs machine learning algorithms, specifically K-Nearest Neighbors (KNN) and Random Forest, to model and predict emotional states based on the extracted features. KNN leverages proximity-based learning, identifying similarities and patterns within the dataset to make accurate classifications. Meanwhile, Random Forest, known for its ensemble learning nature, combines multiple decision trees to offer robust and accurate predictions, further enhancing the system's performance in emotion recognition tasks.

By amalgamating these methodologies, the study achieves an improved accuracy rate in recognizing and categorizing emotions within the DEAP dataset. The hybrid approach of integrating various feature extraction methods, meticulous feature selection, clustering, and machine learning models presents a comprehensive framework for understanding and classifying complex human emotions based on physiological signals. Ultimately, this research contributes valuable insights into the development of robust and accurate emotion recognition systems, laying the groundwork for future advancements in affective computing and human-computer interaction.

## Emotion Recognition From EEG-Based Brain: Computer Interface Using Machine Learning and Deep Learning Approaches

This explores the burgeoning field of human emotion recognition through EEG-based brain–computer interfaces (BCI) and machine learning. Affective com- puting's importance in understanding emotions is emphasized, and traditional methods' limitations are discussed. This study showcases the potential of physiological signals, particularly EEG, for accurate and objective emotion recognition.

In the realm of machine learning, it covers a spectrum of methodologies, including k-nearest neighbor, support vector machine, decision tree, artificial neural network, random forest, naive Bayes, and deep learning techniques like convolutional and recurrent neural networks with long short-term memory (LSTM). Special attention is given to their applications and comparative perfor- mance on various datasets.

The integration of BCIs, specifically EEG, in human–computer interaction sys- tems is addressed, emphasizing the need for emotional intelligence. The study chronicles the evolution of EEG-based BCI emotion recognition techniques, citing their real-time responsiveness and sensitivity to affective state changes.

Human emotion recognition studies published between 2015 and 2021, using EEG data, are scrutinized for insights into the performance of different machines and deep learning algorithms. The text discusses EEG rhythms linked to emotions, revealing the intricate correlation between distinct brain areas and emotional states.

It concludes by highlighting challenges and proposing future research directions in the recognition and classification of human emotional states using EEG. Overall, the paper offers a thorough analysis of the current landscape, advancements, and potential avenues for enhancing emotion recognition through EEG-based BCIs and machine learning.

*Table 3. Models and advantages*

| Reference | Model | Advantages |
|---|---|---|
| (Sidharth et al., 2023) | Transfer Learning in ResNet-50 model | High accuracies in subject-dependent and subject-independent emotion classification |
| (Acharya et al., 2021) | CNN, LSTM | LSTM superior to CNN |
| (Gill & Singh, 2021) | DNN, SVM, KNN, RF, PSD | High accuracy, suitable for real time application |
| (Shao et al., 2019) | DNN, CNN, SGD | CNN's potential for emotional EEG analysis, Highlights the importance of tailored network structures for accuracy. |

continued on following page

*Table 3. Continued*

| Reference | Model | Advantages |
|---|---|---|
| (Bazgir et al., 2018) | SVM | Accuracy less compared to deep learning models |
| (Kusumaningrum et al., 2020) | CNN, SVM, DNN | Superior emotion classification via CNN over traditional DNN approaches |
| (Liu et al., 2016) | Random forest, K-means | Accuracy less compared to deep learning models |
| (Erat et al., 2024) | Machine learning and deep learning models | Deep learning models more accurate than machine learning models |

## CONCLUSION

The investigation of deep learning methods for evaluating EEG data in emotion recognition has demonstrated the considerable promise of these sophisticated computer techniques in improving our comprehension of human emotions. Utilizing different deep learning models, including Convolutional Neural Networks (CNNs), Recurrent Neural Networks (RNNs), Long Short-term Memory networks (LSTMs), and their combinations, has yielded encouraging outcomes in effectively categorizing and forecasting emotional states based on EEG data.

Through this investigation, it is clear that deep learning provides a robust framework for extracting intricate patterns and features from high-dimensional EEG signals, which are indicative of different emotional states. The capacity of these models to acquire knowledge from extensive datasets and discern tiny alterations in brain activity linked to emotions is a helpful instrument for researchers and professionals in neuroscience, psychology, and related disciplines.

Nevertheless, there are still obstacles to overcome in the field of EEG-based emotion recognition. These factors encompass the requirement for extensive and varied datasets to efficiently train models, the computing requirements of handling high-dimensional EEG data, and the imperative for models that can generalize across distinct persons and circumstances. Moreover, the comprehension of deep learning models and their decision-making process is crucial, particularly in contexts where comprehending the model's logic is essential for establishing confidence and ensuring ethical practices, such as in clinical environments.

Possible future research avenues could involve enhancing the resilience and flexibility of deep learning models to effectively process data from diverse populations and contexts, improving the capacity to interpret the models' outcomes, and combining many data sources to enhance the models' comprehension of human emotions. Continued progress in hardware and computational methods will help overcome the existing restrictions, leading to the development of emotion identification systems that are more precise, efficient, and easily accessible.

By further utilizing and improving deep learning methods for EEG analysis, the field can make progress towards creating systems that not only accurately interpret human emotions but also contribute to advancements in mental health treatment, adaptive technologies, and our overall comprehension of the human brain.

## REFERENCES

Acharya, D., Jain, R., Panigrahi, S. S., Sahni, R., Jain, S., Deshmukh, S. P., & Bhardwaj, A. (2021). Multi-class Emotion Classification Using EEG Signals. In Garg, D., Wong, K., Sarangapani, J., & Gupta, S. K. (Eds.), *Advanced Computing* (pp. 474–491). Springer Singapore. DOI: 10.1007/978-981-16-0401-0_38

Adek, R. T., & Ula, M. (2020). A Survey on The Accuracy of Machine Learning Techniques for Intrusion and Anomaly Detection on Public Data Sets. *2020 International Conference on Data Science, Artificial Intelligence, and Business Analytics (DATABIA).* DOI: 10.1109/DATABIA50434.2020.9190436

Alarcão, S. M., & Fonseca, M. J. (2019). Emotions Recognition Using EEG Signals: A Survey. *IEEE Transactions on Affective Computing*, *10*(3), 374–393. DOI: 10.1109/TAFFC.2017.2714671

Baby, K. (2014). *Big Data: An Ultimate Solution in Health Care.*

Bazgir, O., Mohammadi, Z., & Habibi, S. A. H. (2018). Emotion Recognition with Machine Learning Using EEG Signals. *2018 25th National and 3rd International Iranian Conference on Biomedical Engineering (ICBME)*, 1–5. DOI: 10.1109/ICBME.2018.8703559

Craik, A., He, Y., & Contreras-Vidal, J. L. (2019). Deep learning for electroencephalogram (EEG) classification tasks: A review. *Journal of Neural Engineering*, *16*(3), 031001. DOI: 10.1088/1741-2552/ab0ab5 PMID: 30808014

Erat, K., Şahin, E. B., Doğan, F., Merdanoğlu, N., Akcakaya, A., & Durdu, P. O. (2024). Emotion recognition with EEG-based brain-computer interfaces: A systematic literature review. *Multimedia Tools and Applications*, *83*(33), 79647–79694. Advance online publication. DOI: 10.1007/s11042-024-18259-z

Gelenbe, E. (1993). Learning in the Recurrent Random Neural Network. *Neural Computation*, *5*(1), 154–164. DOI: 10.1162/neco.1993.5.1.154

Gill, R., & Singh, J. (2021). Consumer Emotional State Evaluation Using EEG Based Emotion Recognition Using Deep Learning Approach. In Garg, D., Wong, K., Sarangapani, J., & Gupta, S. K. (Eds.), *Advanced Computing* (pp. 113–127). Springer Singapore. DOI: 10.1007/978-981-16-0401-0_9

Hochreiter, S., & Schmidhuber, J. (1997). Long Short-Term Memory. *Neural Computation*, *9*(8), 1735–1780. DOI: 10.1162/neco.1997.9.8.1735 PMID: 9377276

Kusumaningrum, T. D., Faqih, A., & Kusumoputro, B. (2020). Emotion Recognition Based on DEAP Database using EEG Time-Frequency Features and Machine Learning Methods. *Journal of Physics: Conference Series, 1501*, 012020. DOI: 10.1088/1742-6596/1501/1/012020

Lakshminarayanan, V., Ravikumar, A., Sriraman, H., Alla, S., & Chattu, V. K. (2023). Health Care Equity Through Intelligent Edge Computing and Augmented Reality/ Virtual Reality: A Systematic Review. *Journal of Multidisciplinary Healthcare, 16*, 2839–2859. DOI: 10.2147/JMDH.S419923 PMID: 37753339

Liu, J., Meng, H., Nandi, A., & Li, M. (2016). Emotion detection from EEG recordings. *2016 12th International Conference on Natural Computation, Fuzzy Systems and Knowledge Discovery (ICNC-FSKD)*, 1722–1727. DOI: 10.1109/ FSKD.2016.7603437

Ravikumar, A., & Sriraman, H. (2023a). *Acceleration of Image Processing and Computer Vision Algorithms* [Chapter]. Handbook of Research on Computer Vision and Image Processing in the Deep Learning Era; IGI Global. DOI: 10.4018/978-1-7998-8892-5.ch001

Ravikumar, A., & Sriraman, H. (2023b). Mitigating Vanishing Gradient in SGD Optimization in Neural Networks. In Ranganathan, G., Papakostas, G. A., & Rocha, Á. (Eds.), *Inventive Communication and Computational Technologies* (pp. 1–11). Springer Nature Singapore. DOI: 10.1007/978-981-99-5166-6_1

Ravikumar, A., & Sriraman, H. (2024a). A Deep Understanding of Long Short-Term Memory for Solving Vanishing Error Problem: LSTM-VGP. In Baby Maruthi, P., Prasad, S., & Tyagi, A. K. (Eds.), *Machine Learning Algorithms Using Scikit and TensorFlow Environments* (pp. 74–90). IGI Global., DOI: 10.4018/978-1-6684-8531-6.ch004

Ravikumar, A., & Sriraman, H. (2024b). Understanding Convolutional Neural Network With TensorFlow: CNN. In Baby Maruthi, P., Prasad, S., & Tyagi, A. K. (Eds.), *Machine Learning Algorithms Using Scikit and TensorFlow Environments* (pp. 50–73). IGI Global., DOI: 10.4018/978-1-6684-8531-6.ch003

Ravikumar, A., Sriraman, H., Sai Saketh, P. M., Lokesh, S., & Karanam, A. (2022). Effect of neural network structure in accelerating performance and accuracy of a convolutional neural network with GPU/TPU for image analytics. *PeerJ. Computer Science, 8*, e909. DOI: 10.7717/peerj-cs.909 PMID: 35494877

Shao, H.-M., Wang, J.-G., Wang, Y., Yao, Y., & Liu, J. (2019). EEG-Based Emotion Recognition with Deep Convolution Neural Network. *2019 IEEE 8th Data Driven Control and Learning Systems Conference (DDCLS)*, 1225–1229. DOI: 10.1109/DDCLS.2019.8908880

# Chapter 6
# Dynamic SceneNet Spatio-Temporal Interactive Network for Dynamic Video Foreground Detection


**M. Naveen**
*Hindusthan College of Arts and Science, India*

**A. V. Senthil Kumar**
https://orcid.org/0000-0002-8587-7017
*Hindusthan College of Arts and Science, India*

**Ismail Bin Musirin**
*Universiti Teknologi MARA, Malaysia*

**Amit Dutta**
*All India Council for Technical Education, India*

**L. Manjunatha Rao**
*National Assessment and Accreditation Council, India*

**N. Achyutha Prasad**
*East West Institute of Technology, India*

**Indrarini Dyah Irawati**
*Telkom University, Indonesia*

**Giridhar Akula Venkata Shesha**
*Sphoorthy Engineering College, India*

**S. K. Tiwari**
*Dr. C.V. Raman University, India*

**G. Sangeetha**
https://orcid.org/0000-0002-7138-8888
*Christ University, India*

**Binod Kumar**
https://orcid.org/0000-0002-6172-7938
*JSPM's Rajarshi Shahu College of Engineering, Indonesia*

**Ravisankar Malladi**
https://orcid.org/0000-0002-8250-6595
*Koneru Lakshmaiah Education Foundation, India*



DOI: 10.4018/979-8-3693-3840-7.ch006

## ABSTRACT

*This study introduces Dynamic SceneNet, a new Spatio-Temporal Interactive Network developed to tackle the issues of dynamic video foreground recognition. Proposed model uses sophisticated deep learning techniques to learn interactive spatiotemporal characteristics, resulting in a novel solution for accurate and adaptable foreground recognition in dynamic scenarios. Video foreground detection (VFD) is a crucial pre-processing step for accurate target tracking and recognition. Creating a reliable detection network is tough because to interference from shadows, changing backgrounds, and camera jitter. Convolution neural networks have shown to be reliable in several sectors due to their great feature extraction capabilities. This research proposes an interactive spatiotemporal feature learning network (ISFLN) for VFD.. CDnet2014, INO, and AICD datasets corroborate that the introduced ISFLN model surpasses state-of-the-art methodologies in video foreground detection, thanks to its prowess in feature enhancement, interactive multi-scale feature exploration, and deep learning techniques.*

## INTRODUCTION

Foreground recognition in videos, a fundamental aspect of computer vision, has emerged as a revolutionary technology with profound implications across diverse industries. This sophisticated discipline revolves around identifying and isolating foreground elements within video sequences, facilitating a more refined comprehension of visual data. As technology advances, video foreground recognition assumes a pivotal role in numerous applications, spanning from surveillance and security to augmented reality and content generation. Video foreground detection (VFD), a well-explored domain within computer vision, aims at discerning dynamic elements within video sequences and has garnered significant interest across applications such as autonomous driving, remote sensing, action recognition, and video surveillance. As a critical preprocessing step, the precision of foreground detection profoundly impacts the overall quality of subsequent tasks. However, challenges like varying illumination, dynamic backgrounds, shadows, and camera movements intricately complicate the detection process. Over recent decades, diverse methodologies for VFD have been proposed. These approaches can be broadly classified into two main categories: traditional machine learning-based and deep learning-based techniques. Traditional methods, such as pixel-based techniques exemplified by Gaussian Mixture Model (GMM) and Kernel Density Estimation (KDE), encounter notable hurdles in accuracy, especially in the face of illumination shifts and camera motion. Additionally, region- based approaches often lack motion information, while the heightened complexity stemming from multiple block-level computations poses implementation challenges. Traditional methods heavily rely on low-level manual characteristics such as color features, texture features, and spatial distribution, lacking high-level semantic information. This deficiency may result in target missing, detection errors, and a weak response to complex environments.

In recent years, the advent of Convolutional Neural Networks (CNNs) has brought about substantial improvements in various image processing tasks, primarily due to their robust feature extraction capabilities. The utilization of CNNs enables the extraction of high-level semantic cues that may be unattainable through traditional methods. Although many deep learning-based approaches have demonstrated promising results in VFD, there remain certain issues that need to be addressed.

Initially, several existing methods focused solely on analyzing spatial clues, neglecting temporal characteristics and leading to the segregation of information. Secondly, diverse types of features exhibit variations across different levels. In the domain of video detection, incorporating spatio- temporal differences can offer more precise target information. However, some approaches amalgamated these differences during training. Furthermore, certain scholars directly integrated skip connections into the encoder-decoder structure to amplify feature expression. Nevertheless, this approach introduces noise and unnecessary information flow to the decoder, potentially impacting performance.

Addressing the limitations highlighted in existing methods, our approach is motivated by the need to construct a model that maximizes the utilization of spatio-temporal characteristics during the coding phase. Additionally, recognizing the importance of valuable target cues for the decoder, we aim to mine multi-level and multi-scale spatio-temporal features while promoting communication between different types of knowledge. To achieve this, we introduce a two-path spatio-temporal information extraction module (TSIEM) designed to obtain rich spatio-temporal features and reinforce intrinsic connections between them.

A crucial challenge arises in managing the loss of details as information passes through deeper layers in an encoder-decoder network. Our solution to this concern involves proposing a multi- level feature enhancement module (MFEM) instead of a simple skip connection between the encoder and decoder. The MFEM serves the purpose of sharing powerful target information with the decoder, mitigating the potential nuisance caused by the loss of details during information passage.

In summary, this chapter introduces several contributions:

1. The proposal of an innovative end-to-end interactive spatio-temporal feature learning network for video foreground detection. In comparison to existing advanced methods, our model achieves a high detection accuracy while maintaining a faster processing speed of 24 frames per second (fps).

2. The development of a two-path spatio-temporal information extraction module (TSIEM) aimed at capturing multi-level and multi-scale spatio-temporal difference information. The IMFEM enables learning across different feature levels, including low, intermediate, and high levels.
3. Additionally, a multi-level feature enhancement module (MFEM) has been developed to refine coding features for the decoder. This module effectively tackles challenges associated with blurred boundaries and ambiguous pixels caused by coarse features.

## UNDERSTANDING VIDEO FOREGROUND RECOGNITION

### Definition and Basics

Recognizing video foreground involves isolating moving objects or subjects from the background within a video stream. This segmentation process is pivotal for analyzing dynamic scenes, empowering machines to extract pertinent information and make informed decisions. Positioned at the forefront of technological progress, video foreground recognition offers revolutionary solutions across multiple domains. Its applications span from bolstering security and surveillance to transforming our engagement with augmented reality, showcasing its diverse and profound impact. As researchers tackle obstacles and explore new frontiers, the future of video foreground recognition promises further breakthroughs, redefining our understanding and interaction with visual data in the digital era.

### Techniques and Approaches

Video foreground recognition utilizes various techniques, encompassing background subtraction, deep learning methodologies, and optical flow analysis. Background subtraction, a conventional approach, entails juxtaposing individual frames with a reference background to identify alterations. Deep learning techniques, such as convolutional neural networks (CNNs) and recurrent neural networks (RNNs), have emerged as prominent methods owing to their capacity to grasp intricate features and temporal relationships within video data.

## APPLICATIONS OF VIDEO FOREGROUND RECOGNITION

1. **Surveillance and Security**:

The essence of modern surveillance systems lies in the capability of video foreground recognition. By precisely detecting and tracking moving entities, security cameras play a pivotal role in alerting authorities to potential threats, overseeing restricted zones, and bolstering overall safety. Real-time analysis of video feeds enables swift responses and heightened threat identification.

2. Autonomous Vehicles:

In the domain of autonomous vehicles, video foreground recognition stands as a linchpin for discerning pedestrians, fellow vehicles, and road impediments. This technology empowers self- driving vehicles to navigate intricate landscapes, make instantaneous decisions, and uphold the safety of both occupants and pedestrians alike.

3. Augmented Reality (AR) and Virtual Reality (VR):

The integration of video foreground recognition significantly enriches the immersive experiences furnished by AR and VR applications. By accurately isolating and interacting with tangible objects, these technologies seamlessly merge virtual components into the user's surroundings, crafting a more captivating and authentic user interface.

4. Content Creation and Video Editing:

In the sphere of content creation, video foreground recognition streamlines the editing process. Content producers can effortlessly isolate subjects from their surroundings, facilitating the insertion of diverse backgrounds or visual enhancements. This capability holds particular significance in the cinematic realm and for crafting engaging content across platforms like YouTube and TikTok.

## CHALLENGES AND FUTURE DEVELOPMENTS

- **Coping with Complex Scenes and Occlusions:**

Identifying foreground objects faces considerable complexity in scenes featuring multiple moving elements and occlusions. Active research is underway to craft algorithms adept at handling such intricate scenarios with heightened precision.

- **Emphasis on Real-Time Processing**:

The attainment of real-time processing capabilities for video foreground recognition holds paramount importance, particularly in domains like autonomous vehicles and surveillance. Ongoing endeavors focus on refining algorithms and harnessing hardware acceleration to bolster the speed and efficiency of these systems.

- **Addressing Privacy Concerns:**

The escalating usage of video foreground recognition brings forth privacy apprehensions. Striking a delicate balance between the advantages of this technology and safeguarding individual privacy emerges as a pivotal aspect in its deployment.

- **Integration with Emerging Technologies**:

As technological horizons broaden, amalgamating video foreground recognition with other burgeoning technologies such as 5G connectivity, edge computing, and the Internet of Things (IoT) stands to enrich its capabilities and applications further.

Deep learning has transformed video foreground detection, ushering in unparalleled accuracy and efficiency in delineating moving objects from their backgrounds. This potent facet of artificial intelligence has emerged as a cornerstone in the realm of computer vision, furnishing sophisticated solutions for video analysis. In this exploration, we delve into the intricacies of deep learning within video foreground detection, unraveling the techniques, architectures, and applications that underscore its significance.

## DEEP LEARNING TECHNIQUES IN VIDEO FOREGROUND DETECTION

Convolutional Neural Networks (CNNs):

CNNs have demonstrated remarkable effectiveness in capturing spatial features within video frames. In the realm of video foreground detection, CNNs excel in acquiring hierarchical representations of objects and patterns. Utilizing multi-layered convolutional architectures, these networks autonomously extract pertinent features from varying spatial scales, thereby facilitating precise identification of foreground objects.

Recurrent Neural Networks (RNNs):

RNNs possess a proficiency in capturing temporal dependencies inherent in sequential data, rendering them invaluable in video analysis. In the domain of video foreground detection, RNNs proficiently model the temporal evolution of moving objects across frames. This empowers the network to grasp the context of object movements over time, thereby enhancing the overall accuracy of foreground detection, particularly in dynamic scenes.

Long Short-Term Memory (LSTM) Networks:

LSTMs, a specialized variant of RNNs, effectively mitigate the vanishing gradient problem and are adept at modeling long-range dependencies. In the realm of video foreground detection, LSTMs excel in capturing subtle temporal patterns, thereby augmenting the model's capability to discern complex object movements and occlusions.

## ARCHITECTURES FOR VIDEO FOREGROUND DETECTION

Two-Stream Networks:

Two-stream networks amalgamate both spatial and temporal streams to conduct comprehensive video analysis. The spatial stream is dedicated to scrutinizing static features within frames, while the temporal stream encapsulates dynamic changes over time. The fusion of these streams enhances the model's discriminative prowess between foreground and background, particularly in scenarios involving intricate movements.

3D Convolutional Networks:

3D Convolutional Networks extend the functionalities of conventional CNNs by seamlessly integrating the temporal dimension into the convolutional layers. This enables the model to concurrently process spatiotemporal information, rendering it well-suited for video foreground detection tasks where both spatial and temporal features play pivotal roles.

## TRAINING STRATEGIES AND DATA AUGMENTATION

✓ **Transfer Learning:**

Transfer learning leverages pre-trained models on large datasets, such as ImageNet, and fine-tunes them for specific tasks like video foreground detection. This approach accelerates the training process and enhances the model's ability to generalize across diverse video datasets.

**Data Augmentation:**

Augmenting data through techniques like random cropping, flipping, and jittering enhances the resilience of deep learning models by exposing them to a wider array of variations in input data. This is especially advantageous in video foreground detection, where scenes can present diverse lighting conditions, camera angles, and object movements. The advent of deep learning has profoundly transformed video foreground detection, elevating computer vision capabilities to unprecedented levels. By amalgamating advanced neural network architectures, training methodologies, and data augmentation methods, models can now discern and track moving objects with exceptional precision. As technology progresses, deep learning is poised to play an increasingly pivotal role in reshaping the landscape of video analysis, spanning applications from security and autonomous systems to immersive digital experiences. Ongoing research and development in this domain herald a future where video foreground detection becomes an even more integral and seamlessly integrated aspect of our technological ecosystem. Deep learning has revolutionized video foreground detection, furnishing sophisticated solutions for accurately identifying and isolating moving objects within video streams. This paradigm shift from conventional approaches to deep learning techniques has significantly bolstered the accuracy and robustness of foreground detection in dynamic scenarios.

In the realm of video foreground detection, Convolutional Neural Networks (CNNs), in particular, have showcased unparalleled capabilities. Excelling at automatically learning hierarchical features from raw pixel data, CNNs adeptly discern complex patterns and variations in video frames. This intrinsic ability enables deep learning models to adapt and generalize effectively across diverse scenarios, making them well-suited for the challenges posed by dynamic video content. A key advantage of deep learning in video foreground detection lies in its capacity to capture both spatial and temporal features. Unlike traditional methods, which often struggle with temporal dependencies, deep learning models can proficiently analyze the context and motion of objects across consecutive frames. Recurrent Neural Networks (RNNs) and Long Short-Term Memory (LSTM) networks are frequently harnessed to capture temporal dependencies and enhance the overall accuracy of foreground detection. Training deep learning models involves exposing them to extensive labeled video data, facilitating the learning of intricate patterns and correlations between foreground objects and their surroundings. Transfer learning is also commonly employed, leveraging pre-trained models on large datasets for tasks related to video foreground detection, thereby reducing the necessity for extensive labeled data and expediting the training process.

Real-time processing is pivotal in video foreground detection, particularly in applications such as surveillance and autonomous vehicles. Deep learning models can be optimized for deployment on specialized hardware, such as Graphics Processing Units (GPUs) and Field- Programmable Gate Arrays (FPGAs), ensuring swift and efficient processing of video streams. The robustness of deep learning models extends to handling complex scenes with multiple moving objects and occlusions. The hierarchical nature of CNNs enables them to capture both global and local features, facilitating accurate identification of foreground objects even in challenging scenarios. This adaptability proves particularly valuable in applications characterized by dynamic and unpredictable video content. Despite significant progress, challenges persist in the form of potential false positives and negatives. Ongoing research endeavors focus on refining architectures and devising novel loss functions to mitigate these issues. Additionally, the interpretability of deep learning models in video foreground detection remains an area of exploration, as comprehending the decision-making process of these models is pivotal for fostering trust in their applications.

The integration of deep learning with other computer vision techniques further amplifies the capabilities of video foreground detection. Incorporating object tracking, instance segmentation, and 3D scene understanding seamlessly provides a more comprehensive comprehension of video content. This holistic approach enhances the overall performance and applicability of deep learning in video foreground detection. The deployment of deep learning models for video foreground detection transcends traditional domains, finding applications in augmented reality, content creation, and sports analytics. The versatility of these models enables them to contribute to immersive experiences in augmented reality and streamline the content creation process by automating foreground extraction in video editing.

## REVIEW OF LITERATURE

In recent years, notable researchers have conducted experiments exploring various foreground object segmentation techniques. Midhula Vijayan and colleagues introduced a deep neural network architecture that harnesses both temporal and spatial information from background and current processing images. Their experimentation showcased superior performance compared to existing background subtraction methods, demonstrating both qualitative and quantitative advantages.

(Tsung-Han Tsai, 2021) elucidated an unsupervised segmentation technique that utilizes distinct thresholding techniques based on the sorting of changed pixel ratios. This approach facilitates precise decision-making and yields satisfactory results across generic images.

Patrick Dickinson and team addressed foreground segmentation challenges posed by a dynamically changing background over time by employing the Adaptive Gaussian Mixtures model (AGMM). Their experiments surpassed per-pixel and Markov Random Field-based models, achieving a Jaccard coefficient of 0.59.

(Jaime Gallego, 2021) conducted experiments that involved combining pixel-wise and region-based models for foreground segmentation, utilizing one Gaussian per pixel. This approach notably enhanced system performance, particularly for objects with colors similar to the background Jaime Gallego et al. proposed a methodology for monocular static camera sequences and indoor scenarios, utilizing Gaussian pixel color, Modified Mean Shift algorithm, and a Bayesian framework. Their approach demonstrated robust segmentation and tracking capabilities, surpassing state-of-the-art methodologies. Additionally, Jaime Gallego et al. illustrated the reduction of false positives and false negatives by utilizing region-based models for foreground and background region modeling.

(Jiayu Liang, 2018) introduced a novel method incorporating Genetic Programming for feature construction, employing subtree techniques to simultaneously construct multiple features.

(Xuchao Gong, 2021) developed a method using Gaussian Mixture Models (GMM) for modeling static background regions, inter-frame change detection, and Scale-Invariant Feature Transform (SIFT) feature analysis for boundary identification of foreground regions.

(Yizheng Guo, 2020) demonstrated pig segmentation in group-housed environments by combining a mixture of Gaussians using a prediction mechanism and a threshold segmentation algorithm.

(Nikolaos Katsarakis, 2021) enhanced Stauffer and Grimson's algorithm as a baseline by adjusting the learning rate and combining similar Gaussian mixtures, resulting in improved outcomes compared to the baseline algorithm.

In the ever-evolving landscape of artificial intelligence, video foreground detection stands out as a hot topic, with a constant influx of proposed techniques. Our examination and categorization of these approaches revolve around the dual perspectives of traditional methods and deep learning methods.

The inception of popular traditional methods was marked by the Gaussian Mixture Model (GMM). This model operates as a background representation model, leveraging statistical information from pixel samples in video frames. The GMM approach involves constructing a background model by tallying the pixel values for each point in an image, followed by background subtraction to extract moving objects. However, this method is prone to misdetection under certain conditions:

(i) substantial scene changes, such as sudden shifts in light or camera jitter, and
(ii) similarity in colors between the foreground and background.

(Subsequently, Barnich, 2020) introduced a non-parametric method named Vibe, deviating from GMM by adopting a random background update strategy. Given the uncertainty of pixel changes, Vibe employs a random model, acknowledging the suitability of randomness in simulating pixel change uncertainty when a fixed model cannot be determined. Nevertheless, a drawback of the Vibe algorithm is the blending of noise and static targets into the background, causing interference in foreground detection.

To address the dynamic background challenge (Zhao C,2019) pioneered the application of an adaptive threshold segmentation approach. This involved segmenting the input frame into multiple binary images, followed by foreground detection through lateral suppression and an improved template matching method. (Sajid, 2010) proposed a multimode background subtraction (MBS) technique to overcome various challenges. By denoising merged image pixels and creating binary masks in RGB and YCbCr color spaces, this method effectively separated foreground from background.

(Roy, 2020) tackled complex and changing real-world scenarios by constructing three pixel-based background models. Tometal. introduced a video foreground detection algorithm leveraging the spatio-temporal dependency between background and foreground within a tensor framework. These diverse traditional methods showcase attempts to address challenges in foreground detection, with each approach offering unique solutions and coping mechanisms.

Despite the endeavors of these eminent researchers, many proposed models in the literature have struggled to achieve efficient results in foreground segmentation. This paper addresses this gap by focusing on the implementation of a foreground segmentation methodology using deep learning techniques to enhance segmentation results.

As deep learning continues to advance, researchers have integrated this technology into Video Foreground Detection (VFD). (Akula, 2015) applied the LeNet-5 structure for infrared target recognition. (Patil, 2019) innovatively utilized the temporal histogram to estimate the background, sending two saliency maps with different resolutions to a Convolutional Neural Network (CNN) for segmentation. Building on the Fully Convolutional Network (FCN), Yang . introduced dilated convolution to expand the receptive field and proposed a strategy to prevent long-term stationary objects from blending into the background by increasing the interval of multi-frame video sequence images. Guerra et al. employed a U-Net-based background subtraction method to extract the target following background acquisition from a set of video frames.

Recent advancements include the effective application of attention mechanisms in image processing. Minesmatsu. integrated an attention module into a weakly supervised frame- level labeling network to obtain location information. Chen et al. introduced attention mechanisms and residual blocks into Convolutional Long Short-Term Memory (Conv LSTM)

for extracting temporal context cues. Additionally, they incorporated the Spatial Transformer Network (STN) model and Conditional Random Field (CRF) layer at the end of the network for feature refinement. (Qu S, 2021) designed a symmetrical pyramid attention model in CNN to establish close contextual connections.

Other deep learning techniques have been explored. (Sakkos, 2017) employed 3D convolution to simultaneously capture spatial and temporal changes of the target. Akilan . proposed a 3D CNN-LSTM network with dual coding and slow decoding. In the D-DPDL model by Zhao . the convolutional neural network received a random temporal pixel arrangement of features as input, and a Bayesian refinement module was constructed to suppress random noise. Additionally, Bakkay(2018) adopted a conditional generative adversarial network for foreground object detection. Patil. integrated features obtained by the optical flow encoder and edge extraction mechanism into a bridge network composed of dense residual blocks. They propagated the predicted mask of the previous frame to the decoder to obtain precise motion targets. Li. improved detection performance by acquiring and adjusting multi-scale complementary knowledge of the change map through feature extraction, fusion, and refining in three steps.

(Guo, 2012) introduced an object detection approach for tracking objects in video frames. The simulation results demonstrated the effectiveness and accuracy of this technique, showcasing robust performance in the detection of generic object classes. However, there is a need to further enhance classification accuracy, especially in real-time object recognition.

(Ben Ayed, 2015) proposed a method for detecting text data in video frames through big data analytics. This approach involves decomposing video frames into fixed-size blocks and analyzing them using the har wavelet transform technique. The study incorporated a neural network to classify text and non-text blocks. Nevertheless, there is a requirement to focus on extracting regions to eliminate noisy areas and exclude text-like sections.

(Viswanath, 2015) suggested and modeled an approach using non-panoramic background modeling. This method involves representing the entire picture element with one Spatio-temporal Gaussian. The simulation results indicated the ability of this approach to identify moving substances with fewer false alarms. However, the method may fail when adequate features are not obtainable from the section.

(Soundrapandiyan and Mouli, 2015) proposed a novel and adaptive method for pedestrian detection, utilizing image pixel intensities to separate foreground objects from the background. The study achieved a high detection rate of around 90% compared to other single-image methods. Future plans include improving the method's performance with a higher detection rate and lower false positives, aligning with sequence image methods.

(Ramya and Rajeswari, 2016) suggested a modified frame difference method that utilizes the correlation between blocks of the current image and background image to categorize pixels as foreground or background. While the approach showed improvements in finding accuracy with speed, there is a need to focus on utilizing other information available in the blocks, such as shape and edge, to enhance detection accuracy.

(Risha and Kumar, 2016) proposed an optic flow with morphological operations for object detection in videos, emphasizing a static camera setup. The study applied morphological operations to obtain clear moving target images. Future work should address moving cameras and the identification of multiple objects in video frames.

(Najva and Bijoy, 2016) proposed a model for the detection and classification of objects in videos by combining Tensor features with the SIFT approach. This combination was used to classify detected objects using a Deep Neural Network (DNN). Simulation results demonstrated the superiority of the proposed classifier model over existing methods, highlighting its capability to handle large higher-dimensional data with billions of parameters, akin to the human brain.

## EXISTING SYSTEM

This Existing Paper introduces an Artificial Neural Network (ANN) approach that utilizes texture properties to model the background and identify moving objects in video sequences. Extracting foreground objects is a crucial step in information extraction for various computer vision applications. The proposed method exhibits adaptability to dynamic backgrounds. In addition to segmenting moving and non-moving objects in video sequences, detecting moving objects facilitates recognition, classification, and activity analysis.

The approach involves modeling each pixel using Local Binary Pattern (LBP) calculated within a circular region of radius "r" around the pixel. A Background model is constructed using a Self-Organizing Map ANN, and incoming frames are compared against the background model using Mahalanobis distance to determine whether they belong to the background or foreground. The proposed method is designed to handle scenes with moving backgrounds and gradual illumination variations.

To evaluate the algorithm, it has been tested on videos captured by a stationary camera and compared with existing algorithms in terms of detection accuracy and processing speed, crucial factors for color video surveillance systems. The results indicate that our approach achieved a detection accuracy of 93%, and, on average, it is 1.5 times faster than existing algorithms.

## DISADVANTAGES

Nevertheless, employing deep learning in video surveillance comes with two drawbacks. Firstly, there is a limited generalization power. While deep learning outperforms traditional machine learning algorithms, there is room for improvement in its ability to generalize across various scenarios. Domain transfer algorithms have shown some progress in addressing this issue. It is well recognized that utilizing more training data enhances the accuracy of deep learning results. Secondly, deep learning demands a substantial amount of labeled data, particularly at the pixel level in video surveillance. This type of data is more expensive to obtain compared to image classification and detection labels. Despite these challenges, recent efforts have led to the establishment of diverse datasets with pixel-level labeling, catering to the specific requirements of video surveillance scenarios.

## PROPOSED SYSTEM

The Interactive Spatiotemporal Feature Learning Network (ISFLN) primarily consists of two integral components: the Two-Path Spatio-Temporal Information Extraction Module (TSIEM) and the Multi-Level Feature Enhancement Module (MFEM). The subsequent subsections will present a comprehensive analysis of these meticulously designed modules.

## OVERVIEW OF THE PROPOSED SYSTEM

The overall framework of our approach is depicted in Figure 1. Specifically, the Two-Path Spatio-Temporal Information Extraction Module (TSIEM) operates in two stages. In the initial stage, we employ a Siamese Convolutional Network to capture multi-level features from both the current frame and the reference frame. Subsequently, we derive multi-level spatio-temporal difference information through element-wise subtraction. In the second stage, we analyze the spatio-temporal context cues of the object at different scales using an interactive multi-scale feature extraction module. This design offers dual advantages: emphasizing the change information of the object and strengthening the learning of multi-type features from diverse perspectives. Following this, we guide and enhance the original coding features with the Two- Path Spatio-Temporal Information. Finally, this enriched knowledge is shared with the decoder to enhance the expressive capability of the features.

## TWO-PATH SPATIO-TEMPORAL INFORMATION EXTRACTION MODULE (TSIEM)

Many models tend to concatenate the current frame and reference frame directly during feature extraction, overlooking the distinctions among various features. To address this and capture detailed spatio-temporal difference characteristics, we developed a two-path feature extraction module operating at different levels.

For the first path, we input F ∈ RH×W×Ci into the Siamese network to obtain multi- resolution feature maps F1, F2, F3, and F4. Specifically, the number of filters i... (continuation of the original text is incomplete, but you can include the missing portion if needed).

Five convolution operations (Conv_1 → Conv_2 → Conv_3 → Conv_4 → Conv_5) have filter sizes of 32, 64, 128, 256, and 256 (i.e., Ci), respectively. All convolution blocks, except Conv_5, are accompanied by Batch Normalization (BN), Rectified Linear Unit (ReLu), and Max-Pooling. Conv_5, however, only contains BN and ReLu. Subsequently, the absolute difference operation is performed on the corresponding block to obtain the first-path spatio- temporal difference features.

While the aforementioned process captures multi-level features, objects exhibit different scales in diverse scenarios. To address this, we propose an Interactive Multi-Scale Feature Extraction Module (IMFEM) to comprehensively capture information at multiple scales, as depicted in Figure 2.

The IMFEM categorizes features into low-level, intermediate-level, and high-level for processing, involving four steps in the entire learning process. Initially, multi-scale feature extraction operations are applied to low-level features. Typically, two branches of 3 × 3 and 5 ×

5 convolutions are utilized. However, due to computational expense, we substitute 5 × 5 convolution with two 3 × 3 convolutions. Additionally, to reduce the number of channels, a 1 × 1 convolution is added before the 3 × 3 convolution. The branches of 1 × 1 convolution → 3 × 3 convolution are then cross-merged to enhance the exchange of characteristics on the same level. Finally, the fused low-level features at the far end are transmitted to the near end.

*Figure 1. Multi-Level Feature Enhancement Module (MFEM).*

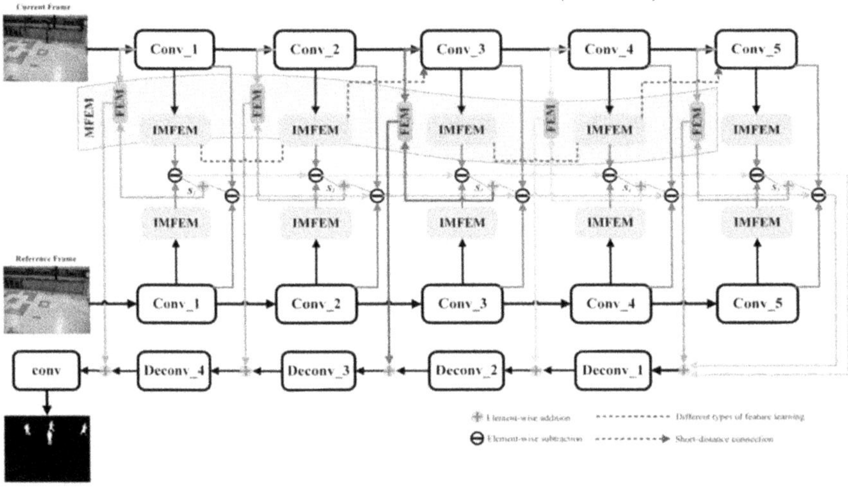


Figure 1 provides an overview of the proposed model, illustrating the Interactive Multi- Scale Feature Extraction Module (IMFEM) and the Multi-Level Feature Enhancement Module (MFEM).

*Figure 2. Interactive Multi-Scale Feature Extraction Module (IMFEM)*

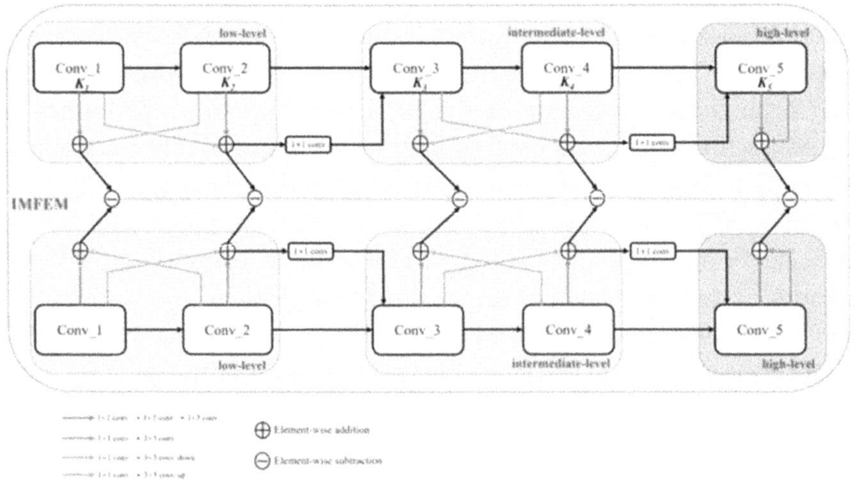


Figure 2 illustrates the Interactive Multi-Scale Feature Extraction Module (IMFEM), where various types of multi-scale information learn from one another.

Intermediate-level features are fused through short-distance connections, where the fused features undergo a 1×1 convolution before connection, reducing the number of channels. The spatio-temporal difference information of corresponding locations is then extracted. The same steps are applied to intermediate-level and high-level features. In the Two-Path Spatio-Temporal Information Extraction Module (TSIEM), one path is utilized for obtaining multi-level features, while the other path acquires multi-scale context features, providing ample target information for the network. The design of the Interactive Multi-Scale Feature Extraction Module (IMFEM) reinforces the learning between the same type and different types of features, promoting the flow of information across levels and enhancing detection effectiveness.

## MULTI-LEVEL FEATURE ENHANCEMENT MODULE (MFEM)

As features progress through deeper convolution layers, some knowledge and details may be lost. To address this, various studies adopt a skip connection approach, directly adding encoding features to the decoder. However, this introduces noise and rough information from the encoding stage, which is not conducive to accurate segmentation of foreground objects. To overcome this, we employ spatio-temporal difference information obtained in the previous section to design a Multi-Level Feature Enhancement Model (MFEM) for enhancing the sharing of encoding features and the decoder, as depicted in Figure 3.

The core of MFEM involves using fused spatio-temporal difference information S{i=1,2,3,4,5} to guide and refine coding features K{m=1,2,3,4,5}. Taking the first feature enhancement module (FEM) as an example, in the initial step, a set of average-pooling and max- pooling operations are performed on S1 and K1 to aggregate spatial information, respectively. The two-path aggregation features are fused using element-wise addition, and the output is passed through a sigmoid activation function to adjust the hybrid features, denoted as M:

[M = sigma{[AP(K1) + MP(K1)] + [AP(S1) + MP(S1)]},]

where AP(•) and MP(•) denote the average-pooling and max-pooling operations, respectively, and $\sigma$ is the sigmoid function.

*Figure 3. Feature Enhancement Module (FEM).*

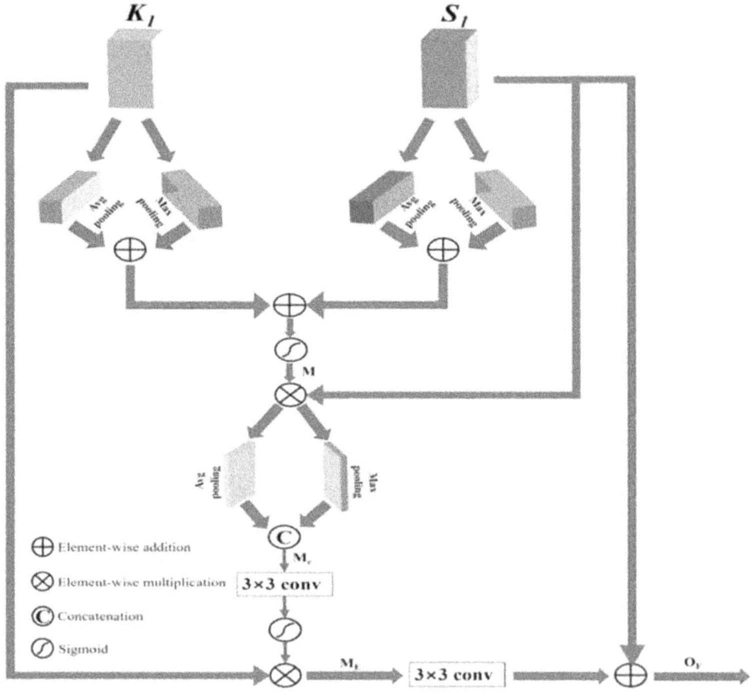


Afterward, the newly obtained features serve as weights and undergo multiplication with S1. Subsequently, the weighted features undergo separate average-pooling and max-pooling operations along the channel axis. This process can be represented by Eq. 2.

[Mc = text{Cat}[AP(S1 otimes M), MP(S1 otimes M)],]
(2)

where (text{Cat}(cdot)) denotes the concatenate operation and (otimes) refers to element-wise multiplication.

Next, the concatenated feature map passes through a 3×3 convolution and sigmoid activation function. The enhanced coding features are obtained through element-wise multiplication with K1. Finally, the enhanced features undergo a 3×3 convolution and are combined with S1 to produce the output OF, which is then transmitted to the corresponding decoder. Similar operations are applied to other levels of coding features. The OF can be formulated using Eqs. 3 and 4.

[ME = sigma(f3 times 3(Mc)) otimes K1, ] (3) [OF = f3 times 3(ME) + S1, ]
(4)

where (f3 times 3(cdot)) represents a 3×3 convolutional layer.

In summary, MFEM utilizes the fused spatio-temporal difference information to guide multi- level coding features, informing them about the importance and location of information. This enhances the expressive ability of coding features and provides valuable clues for the decoder, ensuring a strong guarantee for higher accuracy in detection.

## EXPERIMENTS

### Datasets and Parameter Settings

The video sequences used in the experiments are sourced from LASIESTA [50], CDnet2014 [51], INO [52], and AICD [53] datasets, encompassing indoor and outdoor scenes. During training, 70% of the samples are used as the training set, and the remaining 30% serve as the testing set.

The proposed model is implemented using TensorFlow in Python 3.7 and executed on a workstation with a GeForce RTX 3060 Laptop GPU and i7-10870H CPU. The input frame size is adjusted to 224×224, and the network undergoes 50 epochs with a batch size of 5 for training. The Adam optimizer is employed with an initial learning rate of 0.001. The loss function utilizes binary cross-entropy.

In the experimental analysis, evaluation indicators [45, 51, 54] such as accuracy (Acc), precision (Pre), recall (Rec), F1, percentage of wrong classifications (PWC), false positive rate (FPR), false negative rate (FNR), Specificity (Sp), area under the curve (AUC), and mean intersection over union (mIoU) are utilized.

### Ablation Study

To assess the effectiveness of the proposed modules, a comprehensive analysis is conducted on three datasets (LASIESTA, CDnet2014, INO). Nine indicators are used to observe the performance of the designed modules. Table 1 compares the performance of the one-path spatio- temporal difference module (baseline), Interactive Multi-Scale Feature Extraction Module (IMFEM), and Multi-Level Feature Enhancement Module (MFEM). Specifically, F1 represents the weighted average of precision and recall, and AUC represents the area under the ROC curve, a comprehensive result of False Positive Rate (FPR) and True Positive Rate (TPR, i.e., recall). From the viewpoint of these two composite indicators, our proposed modules are effective. Additionally, Figure 4 presents visual results of different modules. Baseline (BL), TSIEM, and encoder and decoder are directly connected (TSIEM+EDDC) exhibit varying degrees of boundary blur and error detection.

Particularly in the baseline, due to insufficient target information, foreground pixels cannot be correctly judged, resulting in issues such as unclear target contours and missing targets. The addition of IMFEM and MFEM modules sequentially helps alleviate these issues.

*Table 1. The ablation study for different modules.*

| | **Metrics** | | | | | | | | |
|---|---|---|---|---|---|---|---|---|---|
| **Modules** | **Acc↑** | **Pre↑** | **Rec↑** | **F1↑** | **PWC↓** | **FPR↓** | **FNR↓** | **Sp↑** | **AUC ↑** |
| Baseline(BL) | 0.9725 | 0.7848 | 0.7515 | 0.7603 | 1.3545 | 0.0082 | 0.2485 | 0.9918 | 0.9476 |
| BL+IMFEM(TSIEM) | 0.9735 | 0.8324 | 0.8099 | 0.8125 | 1.1616 | 0.0077 | 0.1901 | 0.9923 | 0.9703 |
| TSIEM+EDDC | 0.9749 | 0.8562 | 0.8257 | 0.8338 | 0.8721 | 0.0053 | 0.1743 | 0.9949 | 0.9724 |
| TSIEM+MFEM(our) | 0.9773 | 0.8786 | 0.8547 | 0.8609 | 0.7173 | 0.0042 | 0.1454 | 0.9960 | 0.9739 |

TSIEM+EDDC: Encoder and decoder are directly connected. ↑ Indicates higher value is preferable, ↓ indicates lower value is preferable.

*Figure 4. Visual comparison results of various modules on the LASIESTA dataset. Columns 1 and 2 display the input frame and ground truth, respectively. Columns 3 to 6 showcase the detection results of different modules, respectively.*

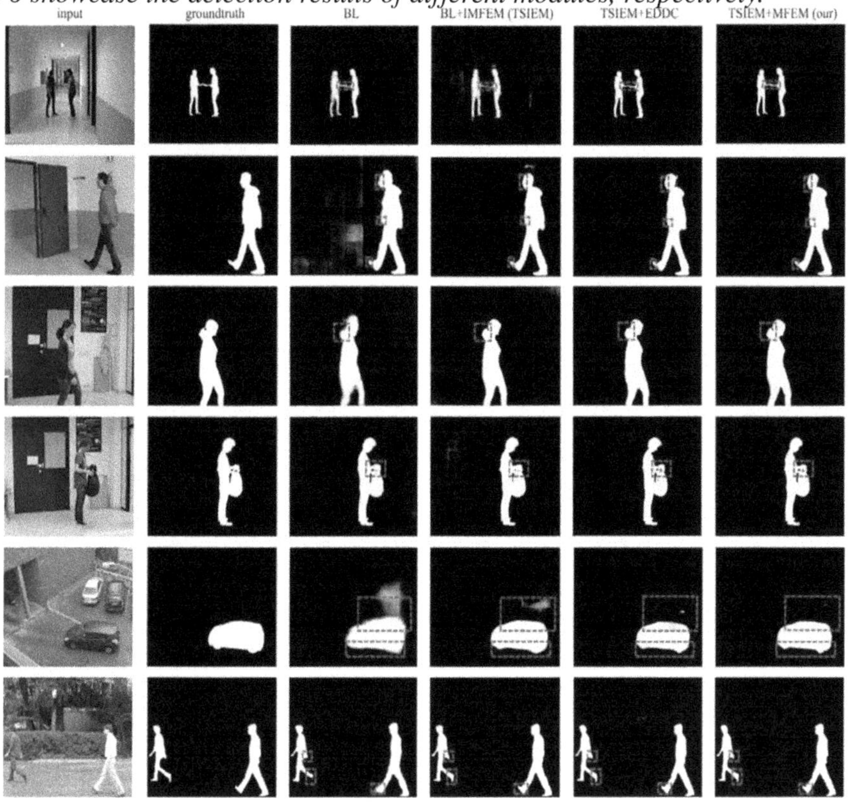


## Comparison With State-of-the-Arts

The performance of ISFLN is compared with existing traditional techniques and deep learning methods across LASIESTA, CDnet2014, INO, and AICD datasets.

1. **LASIESTA dataset:**Table 2 presents the F1 values of various approaches on LASIESTA [16, 55–62]. The last column displays the average F1 value across all videos, which is 89% for the proposed network. In comparison to similar deep learning methods such as FgSegNet-S[16], FgSegNet-M[16], MSFS-51[59], MSFS-55[59], 3CDC-51[60], 3CDC-55[60], and BSUV- Net2.0[61], our technology has demonstrated improvements of 56%, 54%, 53%, 49%, 4%, 5%, and 4%, respectively. Figure 5 visually compares the F1 scores and aver-

age F1 values for different methods. The proposed framework exhibits stable performance across various videos, avoiding abrupt rises or falls.

Moreover, to assess the proposed method's performance on unseen videos, experiments are conducted on the LASIESTA dataset, as detailed in Table 3 and Figure 6. Notably, training and test videos differ. It is essential to highlight that these videos, utilized for experimental purposes, present diverse challenges (refer to Table 3). Both quantitative and qualitative analyses affirm that our approach delivers comparable results for previously unseen videos.

*Table 2. F1 score of various methods on LASIESTA dataset*

| | **Vide os** | | | | | | | | | | |
|---|---|---|---|---|---|---|---|---|---|---|---|
| **Methods** | **IBS** | **ICA** | **IIL** | **IMB** | **IOC** | **ISI** | **OCL** | **ORA** | **OSN** | **OSU** | **Avg** |
| Zovkovik[62]-2004 | 0.53 | 0.83 | 0.24 | 0.87 | *0.95* | 0.91 | 0.88 | 0.83 | 0.38 | 0.71 | 0.71 |
| Maddalena1[55]-2008 | 0.42 | 0.85 | 0.61 | 0.76 | 0.91 | 0.87 | 0.88 | 0.84 | 0.58 | 0.80 | 0.75 |
| Maddalena2[56]-2012 | 0.40 | 0.86 | 0.21 | 0.91 | *0.95* | **0.95** | 0.87 | 0.85 | *0.81* | **0.88** | 0.77 |
| Haines[57]-2014 | 0.68 | *0.89* | 0.85 | 0.84 | 0.92 | 0.89 | 0.83 | 0.86 | 0.17 | *0.86* | 0.78 |
| Cuevas [58]-2018 | 0.66 | 0.84 | 0.65 | *0.93* | 0.78 | 0.88 | *0.93* | 0.87 | 0.78 | 0.72 | 0.81 |
| FgSegNet-S-55[16]- 2018 | 0.22 | 0.60 | 0.39 | 0.60 | 0.23 | 0.39 | 0.23 | 0.15 | 0.13 | 0.37 | 0.33 |
| FgSegNet-M-55[16]- 2018 | 0.21 | 0.69 | 0.32 | 0.71 | 0.31 | 0.43 | 0.22 | 0.18 | 0.19 | 0.25 | 0.35 |
| MSFS-51[59]-2020 | 0.22 | 0.60 | 0.32 | 0.50 | 0.30 | 0.44 | 0.31 | 0.24 | 0.28 | 0.38 | 0.36 |
| MSFS-55[59]-2020 | 0.36 | 0.40 | 0.35 | 0.64 | 0.37 | 0.39 | 0.41 | 0.35 | 0.31 | 0.37 | 0.40 |
| 3CDC-51[60]-2021 | *0.81* | 0.76 | 0.90 | 0.90 | 0.90 | 0.91 | 0.89 | 0.89 | 0.72 | 0.85 | *0.85* |
| 3CDC-55[60]-2021 | 0.72 | 0.82 | *0.92* | 0.89 | 0.91 | 0.87 | 0.87 | *0.90* | 0.69 | 0.85 | 0.84 |
| BSUV-Net2.0[61]- 2021 | 0.77 | 0.68 | 0.88 | 0.81 | **0.96** | *0.92* | *0.93* | **0.94** | **0.84** | 0.79 | *0.85* |
| Our | **0.85** | **0.90** | **0.94** | **0.96** | *0.95* | 0.83 | **0.94** | **0.94** | 0.79 | 0.84 | **0.89** |

The top and second-best results are emphasized in bold and italics, respectively. "Avg" denotes the average F1 score.

*Figure 5. Comparative analysis of F1 and Avg F1 across various methods on LASI-ESTA dataset*

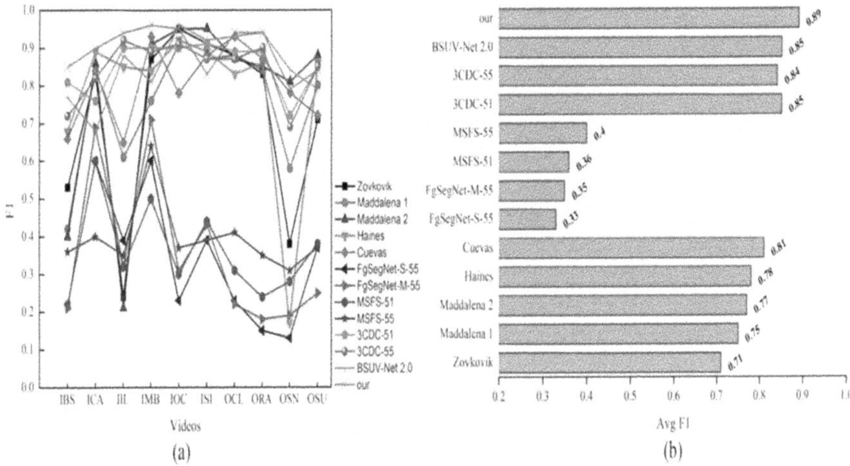


*Table 3. Quantitative analysis for unseen videos on LASIESTA dataset*

| Training | | Testing | | Metrics | | |
|---|---|---|---|---|---|---|
| **Video** | **Challenge** | **Video** | **Challenge** | **Pre↑** | **Rec↑** | **F1↑** |
| I_BS_01 | Bootstrap | I_CA_01 | Camouflage | 0.6035 | 0.7268 | 0.6417 |
| I_IL_01 | Illuminationch ange | I_SI_01 | Simplesequence | 0.9010 | 0.8754 | 0.8805 |
| O_CL_01 | Cloudy | O_SN_01 | Snowy | 0.7926 | 0.7049 | 0.7356 |
| I_CA_02 | Camouflage | I_MB_01 | Modifiedbackgr ound | 0.5637 | 0.6113 | 0.5763 |
| O_CL_01 | Cloudy | O_RA_01 | Rainy | 0.8574 | 0.8265 | 0.8375 |
| I_CA_01 | Camouflage | I_OC_02 | Occlusion | 0.7366 | 0.5031 | 0.5842 |
| O_SN_01 | Snowy | O_RA_01 | Rainy | 0.8572 | 0.7438 | 0.7851 |
| I_IL_01 | Illuminationch ange | I_CA_01 | Camouflage | 0.8914 | 0.5985 | 0.7069 |
| I_IL_02 | Illuminationch ange | I_MB_01 | Modifiedbackgr ound | 0.8483 | 0.9256 | 0.8746 |
| O_CL_02 | Cloudy | O_SU_02 | Sunny | 0.6109 | 0.5264 | 0.5514 |
| Average | | | | 0.7663 | 0.7042 | 0.7174 |

*Figure 6. Visual results of scene independence analysis on LASIESTA dataset*

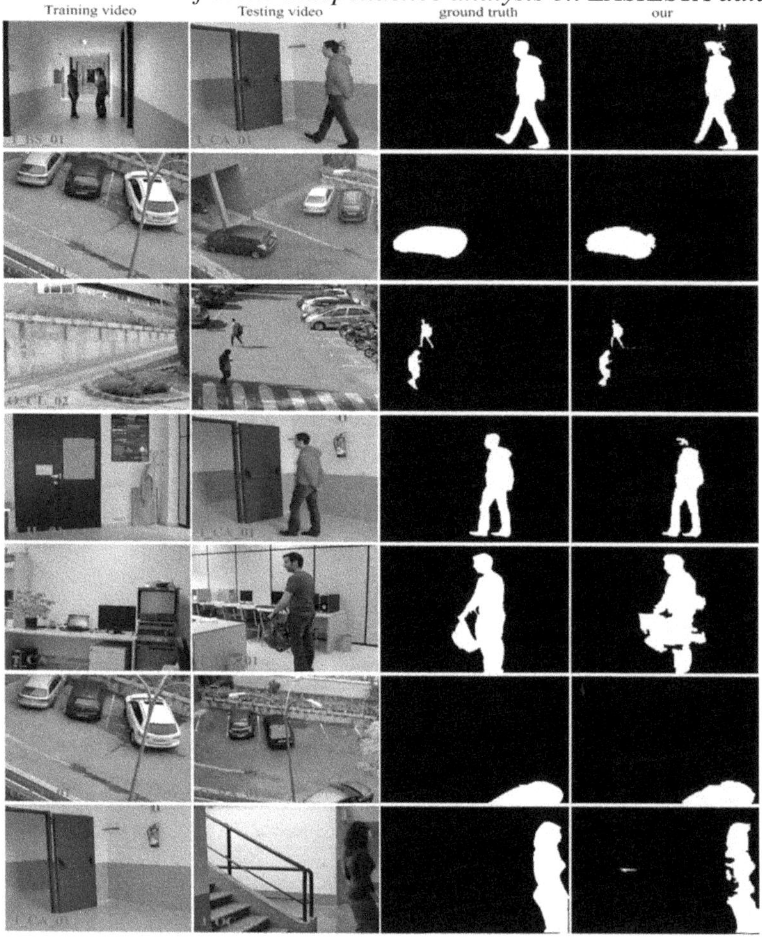


2. CDnet2014 Dataset:

Table 4 showcases the performance of different methods on the CDnet2014 [11, 24, 35, 61–65] dataset, with the F1 value of the proposed network reaching 90.14%. The video used for experimental analysis incorporates numerous challenges, including dynamic background, camera jitter, shadows, and adverse weather conditions. As depicted in Table 4, no single method proves omnipotent. However, in terms of overall performance, our model exhibits greater stability than other methods, a fact further verified in Figure 7. In Figure 7a, the trend of the curve indicates that our proposed approach adeptly handles various challenges.

3. INO Dataset:

Table 5 reports on the qualitative analysis of existing methods [11, 32, 54, 66–68] using accuracy (Acc), recall (Rec), specificity (Sp), and area under the curve (AUC) on the INO dataset. Across various indicators, the proposed method demonstrates a competitive edge. In terms of AUC, the proposed ISFLN achieves 96%, surpassing deep learning techniques like Akula-CNN [32] and SPAMOD [11] by 23% and 6%, respectively.

4. AICD Dataset:

The AICD dataset is employed to assess how effectively the proposed technique handles small targets and significant changes in perspective. Table 5 provides a quantitative comparison with existing advanced methods, utilizing metrics such as F1 and mIoU. As observed in Table 6, the proposed approach outperforms the competition in both metrics. Specifically, when compared with CNN-feat [69], WS-Net [70], DeconvNet [71], Mask-CDNet [72], and SASCNet [45], the F1 of the proposed approach excels.

*Figure 7. Comparative analysis of F1 and AvgF1 across various methods on the CDnet2014 dataset*

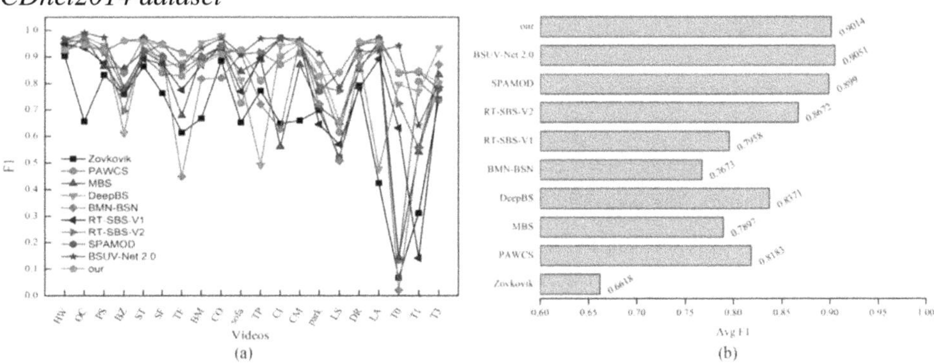


5. Comparative Analysis and Visual Results:

In Table 5, the proposed method demonstrates substantial improvements over existing advanced methods. The F1 metric increases by 60.2%, 74.4%, 54.8%, 51.9%, and 2.2%, while the mIoU increases by 26.6%, 41.2%, 17.6%, 15.2%, and 2.1%, respectively.

6. Visual Results:

Figure 8 offers a qualitative analysis through visual comparison, showcasing a variety of challenges in the displayed videos, including dynamic backgrounds, camera jitter, shadows, turbulence, and adverse weather. The visual results comprise two traditional methods, Zovkovik[62] and PAWCS [63], along with four deep learning methods, DeepBS [64], RT-SBS-V2 [65], SPAMOD[11], and BSUV-Net2.0[61]. Our proposed network consistently produces stable output in the face of various challenges. For instance, in rows 4, 8, 9, 10, and 12, issues like missed detection, holes, and false detection are evident in targets obtained by other approaches. Figure 8 provides visualization results of the proposed method on the AICD dataset. The first row displays the input frame, the second row depicts the ground truth, and the third row presents detection results of the proposed approach. It is evident that the proposed model effectively captures relatively clear objects under small targets and large view variations.

The proposed method relies on the reference frame and the current frame to obtain spatio- temporal difference information for foreground detection, primarily applicable to video surveillance with fixed cameras. However, in experiments, we observe limitations in detection performance when there are significant scene changes and turbulent situations. To address these limitations, future work will focus on guiding the network to identify features by propagating the prediction mask of the previous frame.

6. Real-time Performance:

Time complexity is crucial for evaluating model performance. Table 4 provides a comparison of frames per second (FPS) and the number of trainable parameters for recent methods. Specifically, the proposed network achieves a reasoning speed of 24 frames per Second place, featuring approximately 5.27 million trainable parameters. In summary, our approach demonstrates competitiveness in real-time performance.

*Table 4. Performance scores on INO dataset*

| Methods | Acc↑ | Rec↑ | Sp↑ | AUC ↑ |
|---|---|---|---|---|
| STSM[66]-2015 | 0.75 | 0.70 | 0.28 | 0.70 |
| Akula-CNN[32]-2016 | 0.79 | 0.73 | 0.26 | 0.73 |
| DL[67]-2017 | 0.80 | 0.75 | 0.20 | 0.74 |
| MRF[68]-2018 | 0.81 | 0.79 | 0.19 | 0.78 |
| Qiu[54]-2019 | 0.83 | 0.80 | 0.16 | 0.81 |

continued on following page

*Table 4. Continued*

| Methods | Acc↑ | Rec↑ | Sp↑ | AUC ↑ |
|---|---|---|---|---|
| SPAMOD[11]-2021 | 0.98 | 0.62 | 0.90 | 0.90 |
| Our | 0.98 | 0.77 | 0.99 | 0.96 |

The table 4 presents accuracy (Acc), recall (Rec), specificity (Sp), and area under the curve (AUC) scores for various methods on the INO dataset. The top-performing and second-best results are emphasized in bold and italics, respectively. Avg denotes average F1.

*Figure 8. Qualitative Comparison of Proposed Method and Other Existing Approaches on CDnet2014 Dataset.(Columns 3–9 display the results of Zovkovik[62], PAWCS[63], DeepBS[64], RT-SBS-V2[65], SPAMOD [11], BSUV-Net2.0[61], and our method, respectively.)*

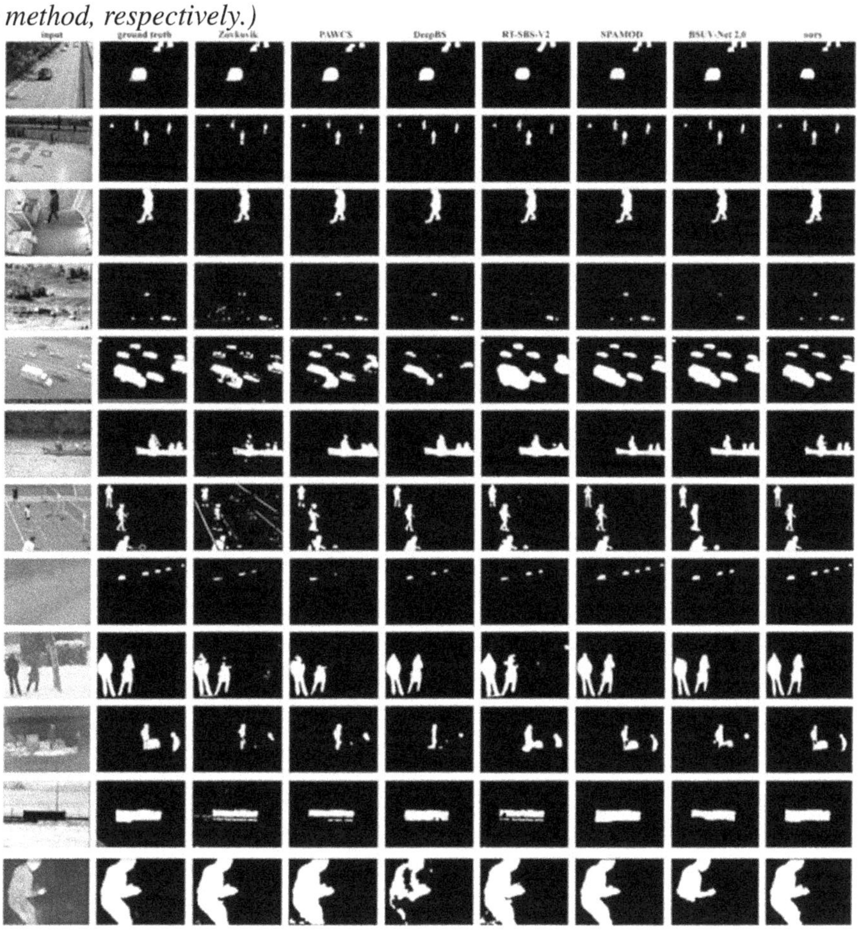

## CONCLUSION

In this study, we introduce a novel end-to-end video foreground detection approach named Interactive Spatio-temporal Feature Learning Network (ISFLN). Our contributions lie in the development of a two-path spatio-temporal information extraction module (TSIEM) and a multi- level feature enhancement module (MFEM). The TSIEM design enhances learning across different information types, acquiring crucial spatio-temporal difference knowledge from multi- level and multi-scale perspectives, vital for effective video object detection. Additionally, MFEM captures fine coding features, shared with the decoder to enhance feature expression, leading to more accurate detection outcomes. The efficacy of the proposed modules is demonstrated through ablation experiments.

Furthermore, when compared to existing advanced approaches across LASIESTA, CDnet2014, INO, and AICD datasets, ISFLN consistently achieves superior detection results while maintaining competitive processing speed (24 fps). In future research, we aim to improve spatial feature learning by incorporating edge label training and delve deeper into scene independence analysis.

## REFERENCES

Ayed, B. (2015). proposed a method for detecting text data in video frames through big data analytics. *Pattern Recognition Letters, 96*, 66–75.

Gallego, J. (2021). Cross-scale feature fusion for object detection in optical remote sensing images. *IEEE Geoscience and Remote Sensing Letters, 18*(3), 431–435. DOI: 10.1109/LGRS.2020.2975541

Gong, X. (2021). Pyramidal graph convolutional network for skeleton-based human action recognition. *IEEE Sensors Journal, 21*(14), 16183–16191. DOI: 10.1109/JSEN.2021.3075722

Guo, Y. (2020). A force levels and gestures integrated multi-task strategy for neural decoding. *Complex & Intelligent Systems, 6*(3), 469–478. DOI: 10.1007/s40747-020-00140-9

Katsarakis, N. (2021). Detection and recognition of stationary vehicles and seatbelts in intelligent Internet of Things traffic management system. *Neural Computing & Applications*. Advance online publication. DOI: 10.1007/s00521-021-05870-6

Liang, J. (2018). Recurrent attention network using spatial- temporal relations for action recognition. *Signal Processing, 145*, 137–145. DOI: 10.1016/j.sigpro.2017.12.008

Najva and Bijoy. (2016). proposed a model for the detection and classification of objects in videos by combining Tensor features with the SIFT approach. *Infrared Physics & Technology, 93*, 205–212.

Qu, S., Zhang, H., Wu, W., Xu, W., & Li, Y. (2021). Symmetric pyramid attention convolutional neural network for moving object detection. *Signal, Image and Video Processing, 15*(8), 1747–1755. DOI: 10.1007/s11760-021-01920-7

Ramya and Rajeswari. (2016). the current image and background image to categorize pixels as foreground or background. *IEEE Trans Intell Transp, 21*(10), 4435–4443.

Risha and Kumar. (2016). proposed an optic flow with morphological operations for object detection. *IEEE Transactions on Image Processing, 20*(6), 1709–1724.

Soundrapandiyan and Mouli. (2015) proposed a novel and adaptive method for pedestrian detection: *Proceedings of the IEEE winter conference on applications of computer vision (WACV)*, pp 2774–2783

Tsai, T.-H. (2021). DFFNet: An IoT-perceptive dual feature fusion network for general real-time semantic segmentation. *Information Sciences*, *565*, 326–343. DOI: 10.1016/j.ins.2021.02.004

Zhao, C. (2019) Non-parametric Model for Background Subtraction. In: *Proceedings of the European conference on computer vision*, pp 751–767

Zhao, C., Sain, A., Qu, Y., Ge, Y., & Hu, H. (2019). Background subtraction based on integration of alternative cues in freely moving camera. *IEEE Transactions on Circuits and Systems for Video Technology*, *29*(7), 1933–1945. DOI: 10.1109/TCSVT.2018.2854273

Chapter 7

# Exploring Brain Imaging Analysis With Nilearn and Related Python Packages

**Kadir Uludag**
https://orcid.org/0000-0003-3713-4670
*Shanghai Jiao Tong University Mental Health Center, China*

## ABSTRACT

*The utilization of brain imaging information holds promise for predicting the prognosis of diverse psychiatric disorders, including schizophrenia, anxiety, attention-deficit hyperactivity disorder, bipolar disorder, and depression. Researchers are displaying growing interest in harnessing Python applications like Nilearn, Nistats, and Nibabel packages to extract and analyze brain imaging data. The Python Nilearn package enables the decoding, analysis, and creation of predictive models using brain imaging data. With Nilearn, users can perform functional connectivity analysis, manipulate brain image volumes, and conduct advanced statistical research on brain images. Furthermore, Nilearn contributes to the development of standardized methods for extracting brain imaging information, addressing potential variations in imaging methodologies across studies. While there is limited literature available on the Nilearn package and similar tools used for similar purposes, it is crucial to promote Nilearn and similar software to facilitate comparisons between studiess.*

## 1. INTRODUCTION

Nilearn, a Python package developed by the nipy community, is a powerful tool for extracting and analyzing brain imaging data. It was created by individuals associated with the INRIA Parietal Project. Nilearn offers various functions that

DOI: 10.4018/979-8-3693-3840-7.ch007

enable the manipulation, processing, and transformation of brain imaging data. It is particularly useful during the preprocessing stage of functional magnetic resonance imaging (fMRI), where essential steps are taken to organize and standardize the data before conducting statistical analysis. Researchers often create preprocessing workflows using Nilearn to streamline these procedures (Esteban et al., 2019). Over time, an increasing number of studies examining the application of fMRI emerged, captivating the interest of many researchers who were intrigued by the potential of analyzing brain imaging data. However, a minority of critics argued that brain imaging tools were no more beneficial than traditional methods in comprehending the nature of neurological and psychiatric disorders.

Furthermore, within the nipy community, there are several additional Python packages such as dipy, mindboggle, nibabel, nitime, popeye, and niwidgets that can enhance brain imaging analysis. These packages contribute to the overall functionality and performance of Nilearn. The objective of our study was to provide an introduction to utilizing the Nilearn Python Package.

## 1.1. Methods

### 1.1.1. Running Sample Nilearn Codes

In order to execute Nilearn package code using Python programming, there are various interfaces available, including Jupyter Lab, Jupyter Notebook, Spyder, and PyCharm. However, if these applications are not accessible, Nilearn samples can still be run using the Mybinder application. Additionally, Jupyter nbviewer allows for the viewing of example codes from the Nilearn package, as mentioned in the official documentation on the Nilearn website (nilearn.github.io). Moreover, human neuroimaging studies typically involve a substantial number of subjects (Cui et al., 2015), and processing neuroimaging data may require heavy software (Huntenburg et al., 2017). Furthermore, utilizing the Nilearn Package necessitates a fundamental understanding of Python programming and familiarity with brain anatomy. Nilearn offers a command-line interface that facilitates the retrieval of atlases (Lawrence et al., 2021). Moreover, the Nilearn package provides the capability to import datasets as illustrative examples. It also offers modules that can be utilized to retrieve sample data for practicing brain imaging processing, such as importing a cortical atlas.

Python's modules consist of bundled classes, objects, functions, and constants employed in various tasks. The installation process involved utilizing pip to install both NumPy and its dependencies, including SciPy, the required libraries (Lemenkova, 2019). We have included seven Nilearn-related code snippets. We have presented a progression of codes, starting from simpler ones and gradually advancing to more complex ones. Subsequently, we have provided a comprehensive explanation of the code. Sample 1 and Sample 2 demonstrate a straightforward approach to loading brain imaging files using Nilearn.

Sample Code 4 showcases the integration of Nilearn to conduct functional analysis. In Sample Code 5, Nilearn is combined with Nistats, while the other sample code focuses on performing fundamental voxel-level analysis utilizing functional MRI data. In the sample code 6, Nilearn was employed for preprocessing purposes. In the tollo3ijy example, Nilearn was combined with scikit-learn (sci-kit learn) for the analysis.

## Sample Code 1:Beginner Nilearn Code

```
from nilearn import datasets, plotting

# Load example functional MRI dataset

dataset = datasets.fetch_adhd(n_subjects=1)

# Retrieve the functional image path

fmri_img = dataset.func[0]

# Plotting the functional image

plotting.plot_img(fmri_img)
```

```
plotting.show()
```

Firstly, to run the code we have to install certain python packages such as nilearn package. The “pip install nilearn” command is used to install the Nilearn package within the Python environment. After installing Nilearn using the “pip install nilearn” command, we can proceed to execute the relevant code. We can also import subpackages using import command.

In this example, the code imports the required modules from Nilearn and fetches an illustrative functional MRI dataset from the ADHD dataset. Next, it obtains the path to the functional image and employs the plotting module to exhibit the functional image. Lastly, the code calls the plotting.show() function to display the plot. In the context of the example code provided, the n_subjects parameter is used to specify the number of subjects or data samples to be fetched from the ADHD dataset. By setting n_subjects=1, the code retrieves data for a single subject from the dataset. The n_subjects parameter allows you to control the amount of data you want to fetch from the dataset, depending on your specific needs or experimentation requirements.

In the example code mentioned, the number `0` is used as an index to access a specific subject or data sample within the fetched dataset. Since `n_subjects` is set to 1, indicating that only one subject's data is being fetched, the index `0` refers to the first (and in this case, the only) subject in the dataset.

## Sample Code 2: Beginner Nilearn Code

```
from nilearn import plotting




# Load a brain atlas image


atlas_img = 'path/to/atlas.nii.gz'




# Display the brain atlas
```

```
plotting.plot_roi(atlas_img, title='Atlas')

plotting.show()
```

In this particular example, the code retrieves a brain atlas image from a designated path, utilizing the variable atlas_img. Subsequently, the function plotting.plot_roi() is employed to exhibit the brain atlas, whereby the title parameter is used to specify the title of the plot. Finally, the plotting.show() function is invoked to display the plot containing the brain atlas image. Furthermore, by importing the plotting module from Nilearn, we gain access to a diverse set of functions that facilitate the visualization of neuroimaging data. We have summarized several commonly employed functions and their respective functionalities:

- plot_img: This function is utilized to present neuroimaging images, such as structural MRI scans or functional MRI time series. It provides customization options, including colormap adjustments, colorbar addition, and plot title specification.
- plot_anat: Specifically designed for visualizing anatomical MRI scans, this function allows you to display structural images while offering customization features like intensity scaling control, overlay addition, and orientation adjustment.
- plot_stat_map: Used for visualizing statistical maps derived from neuroimaging data, such as brain activation maps or statistical contrasts, this function provides options for map thresholding, colormap application, display range control, and colorbar inclusion.
- plot_roi: This function facilitates the display of regions of interest (ROI) defined by binary masks or atlases overlaid on neuroimaging images. It enables the visualization of specific brain regions or areas of interest within the context of the underlying image.

The code referred to as "Title Atlas" serves as the title for the code, and it can be modified or altered as needed.

Also, a NIfTI file, known as a .nii or .nii.gz file, is a widely used for storing complex neuroimaging data. The data can be uploaded to a Jupyter Notebook for future loading and accessing. Furthermore, there is an option to save it as an HTML file.

## Sample Code 3:Use of Numpy with Nilearn Code

```
import numpy as np

from nilearn import datasets, plotting



# Generate a synthetic 3D brain volume

brain_volume = np.random.rand(30, 30, 30)



# Create a Nifti1Image object from the brain volume

brain_img = datasets.niihacks.Nifti1Image(brain_volume, af-
fine=np.eye(4))



# Plotting the brain volume

plotting.plot_img(brain_img, title='Synthetic Brain Volume')

plotting.show()
```

In this particular illustration, the code incorporates NumPy and the essential modules from Nilearn (datasets and plotting). It produces an artificial 3D brain volume by utilizing NumPy's random module. The random module within NumPy provides a diverse range of functions that facilitate the generation of random numbers.

The variable brain_volume is created by using np.random.rand(30, 30, 30), which generates a NumPy array with three dimensions. The size of each dimension, set to 30, represents the width, height, and depth of the brain volume.

Subsequently, the brain volume undergoes conversion into the NIfTI format using the datasets.niihacks.Nifti1Image function. During this conversion, the affine transformation matrix is specified as np.eye(4), which represents an identity matrix. The plotting.plot_img() function is employed to exhibit the brain volume as an image, while the title parameter defines the title of the plot. Lastly, the plotting.show() function is executed to present the plot featuring the synthetic brain volume.

Furthermore, it is possible to create a 4D brain volume. In neuroimaging, a 4D volume commonly represents a sequence of three-dimensional brain volumes acquired over time, commonly referred to as a "time series" or "functional series." Each three-dimensional volume within the 4D dataset corresponds to a specific time point or frame.

To create a 4D brain volume, you would need to generate a NumPy array with four dimensions, where the additional dimension represents the temporal dimension. For example, you can use the np.random.rand() function to generate random values for each voxel in a 4D brain volume:

## Import

```
import numpy as np


brain_volume_4d = np.random.rand(30, 30, 30, 100)
```

In this example, the brain_volume_4d array has dimensions 30x30x30x100, representing a 4D brain volume with a spatial size of 30x30x30 and a temporal dimension of 100 time frames. Each time frame corresponds to a three-dimensional brain volume captured at a specific point in time.

In the context of neuroimaging, the fourth dimension can represent various factors such as distinct subjects, conditions, or experimental parameters, depending on the specific context and structure of the data.

## Sample Code 4: Nilearn to perform functional analysis

```
import numpy as np

from nilearn import datasets, input_data, connectome

# Load the example resting-state fMRI dataset from Nilearn

data = datasets.fetch_development_fmri(n_subjects=1)

fmri_filename = data.func[0]

# Load the brain mask

mask_filename = data.mask_vt[0]

# Load the atlas for defining regions of interest (ROIs)

atlas = datasets.fetch_atlas_harvard_oxford('cort-maxprob-
thr25-2mm')

# Load the time series from the ROIs using a masker

masker = input_data.NiftiLabelsMasker(labels_img=atlas.maps,
standardize=True)

time_series = masker.fit_transform(fmri_filename, con-
founds=None)
```

```
# Perform functional connectivity analysis using correlation

correlation_measure = connectome.ConnectivityMeasure(-
kind='correlation')

correlation_matrix = correlation_measure.fit_transform([-
time_series])[0]

# Print the correlation matrix

print(correlation_matrix)
```

In this specific illustration, we start by loading a sample resting-state fMRI dataset provided by Nilearn. Following that, we acquire the brain mask and an atlas (specifically, the Harvard-Oxford atlas) to delineate ROIs. In the previous code example, the brain mask is used to define the ROI within the brain for functional connectivity analysis. By utilizing the brain mask, the analysis is concentrated solely on brain activity and connectivity, disregarding non-brain parts.

Subsequently, using the NiftiLabelsMasker, we extract the time series data from the ROIs based on the fMRI dataset. Time series data is employed to capture and examine the temporal evolution of a specific variable.

Afterward, we conduct functional connectivity analysis by utilizing correlation as the measure of connectivity. To accomplish this, we employ the ConnectivityMeasure class from Nilearn, which calculates the correlation matrix based on the extracted time series data. Finally, we display the correlation matrix as the outcome of the analysis.

## **Sample Code 5:** Nilearn Package with Nistats

```
import nibabel as nib

from nilearn import plotting
```

```
from nilearn import datasets

from nilearn.input_data import NiftiMasker

from nistats.design_matrix import make_first_level_design_
matrix

from nistats.first_level_model import FirstLevelModel



# Load the brain imaging dataset (for demonstration purpos-
es)

dataset = datasets.fetch_localizer_button_task()



# Load the functional data

func_img = nib.load(dataset.func[0])



# Load the mask data

mask_img = nib.load(dataset.mask)
```

```
# Define the design matrix for first-level analysis

events_file = dataset['events']

design_matrix = make_first_level_design_matrix(events_file)


# Preprocess the functional data using Nilearn

masker = NiftiMasker(mask_img=mask_img, standardize=True)

masked_data = masker.fit_transform(func_img)


# Perform first-level analysis using Nistats

fmri_glm = FirstLevelModel()

fmri_glm = fmri_glm.fit(masked_data, design_matrices=design_
matrix)
```

```
# Generate statistical maps

z_map = fmri_glm.compute_contrast('my_contrast', output_
type='z_score')

t_map = fmri_glm.compute_contrast('my_contrast', output_
type='t_map')

# Visualize the statistical maps using Nilearn

plotting.plot_stat_map(z_map, title='Z-score map', thresh-
old=2.3)

plotting.plot_stat_map(t_map, title='T-score map', thresh-
old=2.3)

# Display the resulting plots

plotting.show()
```

In this illustration, we begin by loading a brain imaging dataset through Nilearn's datasets module. Nilearn's datasets module is a tool that provides access to various neuroimaging datasets. It facilitates working with datasets without the need to load them entirely. Subsequently, we utilize the nibabel library to load the functional and mask data. Following this, we generate a design matrix for the first-level analysis by employing Nistats' make_first_level_design_matrix function.

The functional data undergoes preprocessing using Nilearn's NiftiMasker, and the resultant masked data is provided to Nistats' FirstLevelModel, along with the design matrix. We proceed by fitting the model to the data and computing the statistical maps (z-score and t-score) based on the specified contrast.

The Z-score map reflects the standard score, which measures how many standard deviations a voxel's activation value deviates from the mean. It signifies the statistical significance of the voxel's activity in relation to the average and variability of the data. Larger positive or negative Z-scores indicate more pronounced and significant activations or deactivations, respectively.

On the other hand, the T-score map represents the t-value, which is the ratio of the estimated effect size to its standard error. It quantifies the strength and significance of the activation relative to the noise present in the data.

A threshold of 2.3 mentioned in the code shows a level of significance where the probability, denoted as p, is less than 0.01 in a standard normal distribution.

Ultimately, we employ Nilearn's plotting module to visualize the statistical maps and present the resulting plots.

## Sample Code 6: Basic analysis at the voxel level using functional MRI data

```
from nilearn import datasets, input_data, plotting



# Fetch the example resting-state dataset (ABIDE Pcp)

abide_dataset = datasets.fetch_abide_pcp(data_dir='./data')



# Load the functional MRI data
```

```
fmri_filename = abide_dataset.func_preproc[0]

# Create a masker to extract the brain data

masker = input_data.NiftiMasker(mask_strategy='epi', smooth-
ing_fwhm=6)

masked_data = masker.fit_transform(fmri_filename)

# Perform voxel-wise analysis (e.g., computing mean activa-
tion)

mean_activation = masked_data.mean(axis=0)

# Plot the mean activation as a statistical map

mean_activation_map = masker.inverse_transform(mean_activa-
tion)

plotting.plot_stat_map(mean_activation_map, title='Mean Ac-
tivation')
```

```
# Display the resulting plot

plotting.show()
```

Once the masked data is obtained, we have the opportunity to conduct voxel-level analysis. In this case, we calculate the average activation by computing the mean along the first axis of the masked_data array, encompassing all the voxels.

When the "epi" mask strategy is employed in Nilearn, it generates a binary mask that identifies the brain regions expected to exhibit functional activity, utilizing information from the EPI image.

## Sample Code 7:Nilearn with Preprocessing

```
from nilearn import datasets, plotting, image, input_data



# Fetch the example resting-state dataset (ABIDE Pcp)

abide_dataset = datasets.fetch_abide_pcp(data_dir='./data')



# Load the functional MRI data

fmri_filename = abide_dataset.func_preproc[0]
```

```
# Load the brain mask

mask_filename = abide_dataset.mask

# Apply motion correction to the functional data

motion_corrected_img = image.clean_img(fmri_filename, de-
trend=True, standardize=True)

# Create a masker to extract the brain data

masker = input_data.NiftiMasker(mask_img=mask_filename,
smoothing_fwhm=6)

masked_data = masker.fit_transform(motion_corrected_img)

# Perform voxel-wise analysis (e.g., computing mean activa-
tion)

mean_activation = masked_data.mean(axis=0)
```

```
# Plot the mean activation as a statistical map

mean_activation_map = masker.inverse_transform(mean_activa-
tion)

plotting.plot_stat_map(mean_activation_map, title='Mean Ac-
tivation')

# Display the resulting plot

plotting.show()
```

In this instance, a resting-state functional MRI dataset was obtained from the ABIDE initiative via Nilearn's datasets module and load the functional MRI data from that dataset.

Next, Nilearn's image.clean_img function was applied motion correction to the functional data. This procedure helps mitigate the impact of motion artifacts by detrending and standardizing the data.

Afterward, we load the brain mask from the dataset, which delineates the ROI within the brain.

Following that, a NiftiMasker object is created to extract the brain data. We supply the mask image and specify a smoothing FWHM value of 6mm. The masker is fitted to the motion-corrected functional MRI data, and transform is executed to extract the masked data.

The mean activation is a metric employed to assess the average level of activity or signal strength throughout all voxels.

Following the extraction of the masked data, voxel-wise analysis can be conducted. In this case, we compute the average activation across all voxels by calculating the mean along the first axis of the masked_data array.

## Sample Code 8:Nilearn with Sci-kit

```
from nilearn import datasets, image

from nilearn.input_data import NiftiMasker

from sklearn.model_selection import train_test_split

from sklearn.svm import SVC

from sklearn.metrics import accuracy_score



# Fetch the example motor imagery dataset (Haxby)

haxby_dataset = datasets.fetch_haxby()



# Load the fMRI data and labels

fmri_filename = haxby_dataset.func[0]

labels = haxby_dataset.labels
```

```
# Load the anatomical mask for brain extraction

mask_filename = haxby_dataset.mask



# Create a masker to extract the brain data

masker = NiftiMasker(mask_img=mask_filename, standard-
ize=True)

masked_data = masker.fit_transform(fmri_filename)



# Split the data into training and testing sets

X_train, X_test, y_train, y_test = train_test_split(masked_
data, labels, test_size=0.2, random_state=42)



# Create a support vector classifier

svm = SVC()
```

```
# Train the classifier on the training data

svm.fit(X_train, y_train)



# Predict the labels for the test data

y_pred = svm.predict(X_test)



# Calculate the accuracy of the classifier

accuracy = accuracy_score(y_test, y_pred)

print(f"Accuracy: {accuracy}")
```

In this particular code illustration, we firstly retrieve the Haxby dataset, an fMRI dataset employed for visual object recognition. Both the fMRI data and their corresponding labels are loaded from the dataset.

Later, we load the anatomical mask image, which defines the ROI within the brain.

Subsequently, a NiftiMasker object is created to extract the brain data from the fMRI data. The masker is fitted to the fMRI data, and the transform function is applied to extract the masked data.

Afterwards, the data is divided into training and testing sets using the train_test_split function from scikit-learn.

A support vector classifier is instantiated from scikit-learn. Alternatively, other machine learning methods such as random forest (RF), neural networks or naïve bayes can also be used to run machine learning analysis.

The training data is used to train the classifier by utilizing the fit function, which is a common practice in scikit-learn codes. Predictions for the labels of the test data are made using the trained classifier and the predict function.

Finally, the accuracy of the classifier is calculated by comparing the predicted labels with the true labels using the accuracy_score function from scikit-learn.

This example effectively demonstrates the combination of Nilearn and scikit-learn for machine learning analysis on neuroimaging data. Nilearn is employed for preprocessing and extracting brain data, while scikit-learn is utilized for training and evaluating a support vector classifier. Subsequently, we can also execute a confusion matrix to compute additional accuracy values such as sensitivity and specificity.

## Sample code 9. Regex with Nilearn

```
import re

from nilearn import datasets, image



# Fetch the example motor imagery dataset (Haxby)

haxby_dataset = datasets.fetch_haxby()



# Get the path to the functional MRI data directory

fmri_dir = haxby_dataset.func[0]
```

```
# Define a regular expression pattern to match specific file
names

file_pattern = r'subj\d+_task_motor_run\d+\.nii\.gz'

# Compile the regular expression pattern

regex = re.compile(file_pattern)

# Get a list of file names in the functional MRI data direc-
tory

file_list = image.list_img(fmri_dir)

# Filter the file list using the regular expression pattern

filtered_files = [file for file in file_list if regex.
match(file.filename)]

# Print the filtered file names
```

```
for file in filtered_files:

    print(file.filename)
```

The initial line, "import re," is employed to bring in the regular expression module in Python. Otherwise, we cannot run relevant codes.

In this sample, we firstly acquire the Haxby dataset, which consists of fMRI data loaded previously.

The path to the directory holding the fMRI data is obtained through haxby_dataset.func[0].

Later, a regular expression pattern, file_pattern, is defined using r'subj\d+_task_motor_run\d+.nii.gz'. This pattern is tailored to match file names adhering to a specific format, such as subj1_task_motor_run1.nii.gz, subj2_task_motor_run2.nii.gz, and so forth.

The code r'subj\d+_task_motor_run\d+\.nii\.gz' represents a regular expression pattern. We can explain the code as follows:

- r: The 'r' prefix before the string indicates a raw string in Python, which treats backslashes () as literal characters rather than escape characters.
- subj: This part of the pattern is matched exactly as it is, representing the word "subj" in the file name.
- \d+: The \d is a special character in regex that matches any digit (0-9). The + following \d means that one or more digits can occur after the "subj" part of the file name.
- _task_motor_run: This part of the pattern is matched exactly as it is, representing the phrase "_task_motor_run" in the file name.
- \d+: Similar to the previous occurrence of \d+, this matches one or more digits that can occur after the "_task_motor_run" part of the file name.
- \.nii\.gz: The \. matches a literal dot (.), and nii\.gz matches the characters "nii.gz" exactly. Together, they represent the file extension ".nii.gz".

Similar codes can helps improving regular expression skills and it can be used for various reasons. The regular expression pattern is compiled into a regex object using re.compile() and assigned to regex. By employing image.list_img(), a list of file names within the functional MRI data directory is retrieved.

The file list is filtered by iterating over each file and verifying if it matches the regular expression pattern using regex.match(file.filename).

The files that match the pattern are accumulated in the filtered_files list.

Finally, the filenames of the filtered files are printed.

Overall, This code example effectively illustrates how Nilearn can use regular expressions to selectively choose specific files based on their names or certain patterns for subsequent analysis or processing.

## 2.1.Plotting Images Using Nilearn

Nilearn organizes its data visualization tools within the plotting module. Similar to the functionality of the Matplotlib package, Nilearn's plotting module enables the creation of intricate graphics for brain imaging. It supports the plotting of various types of images, including anatomical images, T2 images, statistical maps, masks, connectomes, network nodes, atlas maps, and voxel intensities. Nilearn assumes that brain imaging data is stored in nifti images. With the help of Nilearn's plotting function, we can generate visualizations of diverse shapes, such as seed-to-voxel correlation maps and anatomical images.

In this manuscript, we provided illustrations of nilearn plotting through various examples (e.g. sample 1).

## 2.2. Statistical Analysis and Nilearn

Neuroimaging data offers various possibilities for applications such as decoding, predictive modeling, functional connectivity analysis, brain parcellations, and connectome analysis. The statistical functionalities in Nilearn are housed within the `glm` module, which provides tools to analyze brain data and quantify the brain's functioning through algorithms.

## 2.3. Nilearn and Brain Connectivity

The examination of intricate networks enables the characterization of the brain in terms of a network structure (Magalhães et al., 2015). Nilearn offers the capability to perform comprehensive analysis of brain connectivity, as demonstrated in the provided Nilearn samples (nilearn.github.io).

## 2.4. Preprocessing Related to Brain Imaging and Nilearn Package

FMRI artifacts can arise from various sources, including head motion, physiological factors, motion, and physiological sources, leading to potential distortions or corrupting effects (Jo et al., 2013). Artifacts in fMRI data, related to motion and physiological sources impact the signal and noise (Kelly et al., 2010). As a result, it becomes essential to preprocess the pertinent brain images.

The Python Nilearn package offers various preprocessing capabilities, including applying masks, realignment, and classes for masking or smoothing complex data. However, there is no specific rule dictating the order in which preprocessing steps should be performed (Parker and Razlighi, 2019), and the order may change according to the goal and design of the study.

To illustrate, the realignment step in preprocessing involves the correction of brain motion and the alignment of images in a temporal sequence (Jaber et al., 2019). In fMRI, slices are not acquired simultaneously and can experience misalignment. To address this issue, preprocessing pipelines include timing correction methods (Parker and Razlighi, 2019). Furthermore, slice timing correction is susceptible to the influence of image processing techniques that encompass multiple slices, with one common example being 3D smoothing (Parker and Razlighi, 2019).

## 2.5. Python Packages Used to Support Nilearn Package

Several commonly utilized Python packages provide support for Nilearn. Familiarizing ourselves with their functionalities aids in comprehending how the Nilearn package operates. Python packages such as numpy, scikit-learn, DICOM, Nibabel, and Matplotlib have been employed to bolster the Nilearn package's capabilities.

### 2.5.1. Numpy Package of Python and Nilearn Package

NumPy provides a flexible array structure that aids in executing specific analyses. Numpy arrays serve as the storage structure for matrices, and Nilearn can generate data using Numpy functions. As tuples are immutable and cannot be modified directly, they need to be converted into Numpy arrays to enable alterations. Therefore, the Numpy package is frequently employed in conjunction with Nilearn.

NumPy serves as the predominant array programming library in Python, playing a crucial role in research analysis pipelines across a wide range of fields, including physics, chemistry, astronomy, geoscience, biology, psychology, materials science, engineering, finance, and economics (Harris et al., 2020). It includes several subpackages such as core, lib, testing, and random (Oliphant, 2006). NumPy offers an extensive collection of mathematical functions that can be applied to the ndarray object and these functions encompass a variety of mathematical operations, ranging from basic algebraic functions like addition and multiplication to trigonometric functions such as sine, cosine, and others (Oliphant, 2006).

Moreover, examples were provided to illustrate the utilization of NumPy in conjunction with Nilearn (example 3).

### 2.5.2. Sci-kit Learn Package of Python and Use of Nilearn

Sci kit learn package of Python (Pedregosa et al., 2011) is implemented to solve complex classification or regression problems in brain imaging literature using various ML models such as support vector machine (SVM), RF, or naïve bayes (NB). In research focused on psychiatric prediction models, these models have found common application in predicting treatment outcomes for psychiatric disorders such as schizophrenia (Uludag et al., 2023), bipolar disorder, and depression.

To evaluate the performance of machine learning models, metrics such as ROC (receiver operating characteristics), AUC (area under curve), confusion matrix, F1 score, and accuracy values from sci-kit learn are commonly employed.

When it comes to reusing ML models, saving them becomes essential. This can be achieved using the Joblib or Pickle packages in Python. In the realm of brain imaging, Sci-kit Learn can also be utilized to differentiate between two categories of images. Sample code 8 mentioned in the manuscript includes sci-kit learn code with Nilearn.

### 2.5.3. DICOM and Niftii Images and Nilearn

To load the installed DICOM Python package, one needs to execute the "Import DICOM" code. This package adheres to the standard for digital images and provides a file format for exchanging images (Riddle and Pickens, 2005). A DICOM image represents a 2D image as it typically consists of a single slice from an MRI or CT scan. The NIFTI file can contain a set of these DICOM images.

### 2.5.4. Scikit Image

The scikit-image package, similar to Python's Pillow package, is employed for image processing tasks. It can be imported using the "import skimage" code. This package offers a range of functionalities, including accessing pixel values, rotating, shifting, scaling, and drawing images.

### 2.5.5.Matplotlib Package

Matplotlib is a frequently utilized Python package for the purpose of creating visual plots and graphics (Ari and Ustazhanov, 2014). Matplotlib can be used from the Python shell (Barrett et al., 2005). Matplotlib is widely employed to generate diverse types of visualizations, including boxplots, bar graphs, and histograms, without the need for intricate coding. It is also utilized to support the sci-kit learn package in Python, enabling the creation of various graphics such as ROC-AUC curves for ML analysis while allowing for convenient saving of the generated plots.

### 2.5.6. Scipy Python Package

Scipy is an adaptable package that expands upon the fundamental functionalities of SciPy and upholds rigorous standards for implementation (Virtanen et al., 2020). Scipy is utilized within Nilearn studies to perform correlation tests like the Pearson product-moment correlation, Spearman correlation, or Kendall-tau correlation. Additionally, Scipy is commonly employed for tasks such as computational geometry, signal processing, and implementing filters. It also provides the capability to apply the Kolmogorov-Smirnov test, although it should be noted that this approach can be subject to controversy (Aguirre et al., 1998). Overall, the Scipy package serves as a valuable tool in supporting the analysis conducted using Nilearn.

### 2.5.7. Regular expressions

Regular expressions (REGEX) provide the capability to search, filter, and manipulate text and numeric data, offering a powerful tool for these operations (Yarberry, 2021). It can help to preprocess the brain imaging data and its functions. Code Sample 9 showcased an example that utilized regex.

### 2.5.8.Nibabel Python Package

Nibabel is primarily utilized for accessing various neuroimaging file formats and can effectively support the Nilearn package. To import the package, one can use the code "import nibabel as nib". Code Sample 5 demonstrated an example that involved the utilization of nibabel code.

### 2.5.9. Niwidgets

Niwidgets, a Python package, was originally created by Bjoern Soergel and Jan Freyberg, and it received subsequent enhancements from volunteer developers. To install the Niwidgets Python package, the "pip install 'niwidgets'" code was utilized. The repository offers convenient wrappers that enable the display of interactive widgets for visualizing neuroimaging data (https://github.com/nipy/niwidgets). By using the capabilities offered by Niwidgets, individuals involved in complex neuroimaging research and practice can improve their data analysis processes by integrating interactive visualizations and widgets into their Jupyter notebooks.

#### 2.5.9.1. Tensorflow Python Package

TensorFlow implements a probability theory that has been adjusted to align with the deep-learning approach of differentiable computation (Dillon et al., 2017). TensorFlow is a deep learning framework known for its relatively high speed and efficiency (Prakash et al., 2021). TensorFlow presents a multitude of possibilities in the field of biological research, providing researchers with a standardized system that has the potential to enhance usability and convenience (Rampasek and Goldenberg, 2016). It can help to create prediction models using brain imaging data. Scikit-learn can also be used for similar goals. However, tensorflow mainly related to deep learning task. Scikit-learn is commonly used in the various fields for conducting machine learning analysis such as psychiatry.

#### 2.5.9.2. Nitime Package

To utilize the Nitime package, you need to execute the code "pip install nitime" in order to install it. The package encompasses algorithms designed for time-series analysis in both the time and spectral domains.

#### 2.5.9.3. OS package of Python

This module offers a means to utilize functionality that is specific to the operating system being used. Its primary purpose is to facilitate operations related to directories.

#### 2.5.9.4. Pillow Package of Python

The image library is specifically created to access and manipulate data stored in pixel form. It serves as a fundamental component for a versatile image processing tool. In essence, this package enables various operations such as opening, rotating, displaying, merging, filtering, and saving images related to brain imaging. Consequently, it can be a valuable asset in complementing the functionalities of the Nilearn package.

#### 2.5.9.5. OpenCV Package of Python for Brain Imaging

OpenCV is a library primarily employed for the purpose of analyzing images (Culjak et al., 2012). It finds utility in various applications, such as medical image analysis, video processing, facial recognition, object tracking, and 3D model extraction (Guillen, 2019). It offers versatility in achieving multiple objectives, such as blending two images, enhancing image contrast, applying image smoothing techniques, performing image thresholding, and computing gradients to detect edges.

#### 2.5.9.6. Time Package of Python

The Python time module offers a range of functions related to process handling. Additionally, the datetime and calendar modules provide additional functionality in this domain. The time package is commonly employed in various brain imaging problems that require relevant time-related operations.

#### 2.5.9.7. Pandas Package of Python

Python incorporates a data analysis tool known as the pandas package, which offers extensive capabilities for data manipulation and management (Bernard, 2016). Pandas allows users to perform transformations, filtering, and modifications on the DataFrame, as it is a versatile framework (Hagedorn et al., 2021). PANDA has the ability to streamline the image processing of dMRI datasets and facilitate research on human connectomes (Cui et al., 2013).

#### 2.5.9.8. Nistats

There is limited existing literature on the application of Nistats. However, this tool proves useful in extracting predicted time series, which aids in evaluating the model fit quality. Similar to the Nilearn package in Python, the Nistats package also offers dataset samples that can be accessed as illustrative examples. Nistats supports several functionalities such as general linear model.

## 2.6.Brain Imaging Problems and Nilearn

A prior research study highlighted issues concerning statistical inference and sample size in the context of MRI methods (Hupé, 2015). Moreover, the comparisons made in functional connectivity studies can exhibit inconsistencies (Horwitz, 2003). Furthermore, journal staff members have expressed the need for fMRI workshops and guidelines pertaining to the conduction of fMRI research (Weber et al., 2015).

### 2.6.1. Problems Related to Installing Nilearn

The installation process for Nilearn may vary depending on the development environment employed, such as Jupyter Lab, Jupyter Notebook, or Spyder. Furthermore, using an outdated version of Nilearn or Python can potentially lead to technical problems that hinder the execution of the code. We have mentioned several issues.

Dependency Issues: Nilearn has dependencies on other packages such as NumPy, SciPy, and scikit-learn. Ensure that these dependencies are installed and up to date before attempting to install Nilearn. You can use a package manager like pip to install or update these dependencies.

Python Version: Nilearn may require a specific version of Python.

Internet Connectivity can also cause problems. If you are encountering installation errors related to network connectivity, ensure that you have a stable internet connection. In some cases, using a different network or configuring proxy settings might resolve the issue. Additionally, the utilization of a VPN can result in potential issues or complications.

### 2.6.2. Memory Problems Associated With Nilearn

Memory consumption-related issues can arise, impeding the proper execution of Python codes. To address this problem, various solutions can be devised, such as simplifying complex layers or reducing their complexity.

### 2.6.3. Cython

Cython serves as a static compiler that optimizes code for the Python programming language. Cython, the Python C extension, to incorporate specific functions that cannot be executed using the vectorized capabilities of the NumPy library (Morra and Morra, 2018). By incorporating programs written in C, we can enhance the performance and efficiency of brain imaging code.

## 2.7. Limitations

This manuscript lacks the inclusion of essential and complex Python codes required for conducting brain imaging analysis. We have provided only a limited number of code examples that are relevant to brain imaging. Furthermore, it does not serve as a comprehensive systematic review that encompasses Nilearn codes. The lack of relevant Python codes hampers the practical application of the proposed techniques in real-world scenarios. The absence of Python codes deprives readers of crucial insights and specific information regarding implementation decisions, parameter configurations, and potential challenges encountered throughout the brain imaging analysis process.

## 2.8 Suggestions for Further Studies

The codebase of Nilearn is subject to potential modifications and enhancements, leading to the addition of more functionality. Consequently, future studies may incorporate these changes into the Nilearn program. Additionally, it is recommended that future studies provide a detailed description of the coding analysis employed to conduct an fMRI study using Nilearn. In addition, educational videos can be leveraged as a valuable resource for teaching the Nilearn package. Educational videos also provide the benefit of visualizing the code's output. Furthermore, chatbots can be used to run and analyze nilearn and complex brain imaging codes. As chatbots are capable of assisting in error identification and correction, they can serve as dynamic tools for code writing purposes.

## CONFLICT OF INTEREST STATEMENT

There is no conflict of interest to declare.


## ACKNOWLEDGMENTS

Thanks to the Nilearn Developers who have created the Nilearn Package.

## FUNDING

There is no specific funding to conduct this study.


## DATA AVAILABILITY STATEMENT

There is no associated data. We have included the relevant code within the chapter for reference.

## REFERENCES


Aguirre, G. K., & Zarahn, E., & D′ Esposito, M. (1998). A critique of the use of the Kolmogorov-Smirnov (KS) statistic for the analysis of BOLD fMRI data. *Magnetic Resonance in Medicine*, *39*(3), 500–505.

Ari, N., & Ustazhanov, M. (2014). Matplotlib in python. 2014 11th International Conference on Electronics, Computer and Computation (ICECCO),. IEEE, 1-6.

Ascher, D., Dubois, P. F., Hinsen, K., Hugunin, J., & Oliphant, T. (2001). Numerical python.

Barrett, P., Hunter, J., Miller, J. T., Hsu, J.-C. & Greenfield, P. matplotlib--A Portable Python Plotting Package. Astronomical data analysis software and systems XIV, 2005. 91.

Bernard, J. (2016). *Python data analysis with pandas. Python Recipes Handbook*. Springer. DOI: 10.1007/978-1-4842-0241-8

Cui, Z., Zhong, S., Xu, P., Gong, G., & He, Y. (2013). PANDA: A pipeline toolbox for analyzing brain diffusion images. *Frontiers in Human Neuroscience*, *7*, 42. DOI: 10.3389/fnhum.2013.00042 PMID: 23439846

Culjak, I., Abram, D., Pribanic, T., Dzapo, H., & Cifrek, M. (2012, May). A brief introduction to OpenCV. In 2012 proceedings of the 35th international convention MIPRO (pp. 1725-1730). IEEE.

Daily, J., & Lewis, R. R. (2011, September). Using the Global Arrays Toolkit to Reimplement NumPy for Distributed Computation. In SciPy (pp. 29-35).

Dillon, J. V., Langmore, I., Tran, D., Brevdo, E., Vasudevan, S., Moore, D., Patton, B., Alemi, A., Hoffman, M., & Saurous, R. A. (2017). Tensorflow distributions. *arXiv preprint arXiv:1711.10604*.

Esteban, O., Markiewicz, C. J., Blair, R. W., Moodie, C. A., Isik, A. I., Erramuzpe, A., Kent, J. D., Goncalves, M., Dupre, E., Snyder, M., Oya, H., Ghosh, S. S., Wright, J., Durnez, J., Poldrack, R. A., & Gorgolewski, K. J. (2019). fMRIPrep: A robust preprocessing pipeline for functional MRI. *Nature Methods*, *16*(1), 111–116. DOI: 10.1038/s41592-018-0235-4 PMID: 30532080

Guillen, G. (2019). *Digital Image Processing with Python and OpenCV. Sensor Projects with Raspberry Pi*. Springer. DOI: 10.1007/978-1-4842-5299-4

Hagedorn, S., Kläbe, S., & Sattler, K.-U. (2021). *Putting Pandas in a Box*. CIDR.

Harris, C. R., Millman, K. J., Van Der Walt, S. J., Gommers, R., Virtanen, P., Cournapeau, D., Wieser, E., Taylor, J., Berg, S., Smith, N. J., Kern, R., Picus, M., Hoyer, S., Van Kerkwijk, M. H., Brett, M., Haldane, A., Del Río, J. F., Wiebe, M., Peterson, P., & Oliphant, T. E. (2020). Array programming with NumPy. *Nature*, *585*(7825), 357–362. DOI: 10.1038/s41586-020-2649-2 PMID: 32939066

Horwitz, B. (2003). The elusive concept of brain connectivity. *NeuroImage*, *19*(2), 466–470. DOI: 10.1016/S1053-8119(03)00112-5 PMID: 12814595

Huntenburg, J., Abraham, A., Loula, J., Liem, F., Dadi, K., & Varoquaux, G. (2017). Loading and plotting of cortical surface representations in Nilearn. *Research Ideas and Outcomes*, *3*, e12342. DOI: 10.3897/rio.3.e12342

Hupé, J.-M. (2015). Statistical inferences under the Null hypothesis: Common mistakes and pitfalls in neuroimaging studies. *Frontiers in Neuroscience*, •••, 9. PMID: 25745383

Jaber, H. A., Aljobouri, H. K., Çankaya, I., Koçak, O. M., & Algin, O. (2019). Preparing fmri data for postprocessing: Conversion modalities, preprocessing pipeline, and parametric and nonparametric approaches. *IEEE Access : Practical Innovations, Open Solutions*, *7*, 122864–122877. DOI: 10.1109/ACCESS.2019.2937482

Jo, H. J., Gotts, S. J., Reynolds, R. C., Bandettini, P. A., Martin, A., Cox, R. W., & Saad, Z. S. (2013). Effective Preprocessing Procedures Virtually Eliminate Distance-Dependent Motion Artifacts in Resting State FMRI. *Journal of Applied Mathematics*, *2013*, 935154. DOI: 10.1155/2013/935154 PMID: 24415902

Kelly, R. E.Jr, Alexopoulos, G. S., Wang, Z., Gunning, F. M., Murphy, C. F., Morimoto, S. S., Kanellopoulos, D., Jia, Z., Lim, K. O., & Hoptman, M. J. (2010). Visual inspection of independent components: Defining a procedure for artifact removal from fMRI data. *Journal of Neuroscience Methods*, *189*(2), 233–245. DOI: 10.1016/j.jneumeth.2010.03.028 PMID: 20381530

Lawrence, R. M., Bridgeford, E. W., Myers, P. E., Arvapalli, G. C., Ramachandran, S. C., Pisner, D. A., Frank, P. F., Lemmer, A. D., Nikolaidis, A., & Vogelstein, J. T. (2021). Standardizing human brain parcellations. *Scientific Data*, *8*(1), 78. DOI: 10.1038/s41597-021-00849-3 PMID: 33686079

Lemenkova, P. (2019). Processing oceanographic data by Python libraries NumPy, SciPy and Pandas. *Aquatic Research*, *2*, 73–91. DOI: 10.3153/AR19009

Magalhães, R., Marques, P., Veloso, T., Soares, J. M., Sousa, N., & Alves, V. (2015). Construction of functional brain connectivity networks. In Distributed Computing and Artificial Intelligence, 12th International Conference (pp. 303-310). Springer International Publishing.

Morra, G. & Morra, G. (2018). Fast Python: NumPy and Cython. *Pythonic Geodynamics: Implementations for Fast Computing*, 35-60.

Oliphant, T. E. (2006). *Guide to numpy*. Trelgol Publishing USA.

Parker, D. B., & Razlighi, Q. R. (2019). The Benefit of Slice Timing Correction in Common fMRI Preprocessing Pipelines. *Frontiers in Neuroscience*, *13*, 13. DOI: 10.3389/fnins.2019.00821 PMID: 31551667

Pedregosa, F., Varoquaux, G., Gramfort, A., Michel, V., Thirion, B., Grisel, O., Blondel, M., Prettenhofer, P., Weiss, R. & Dubourg, V. (2011). Scikit-learn: Machine learning in Python. *the Journal of machine Learning research,* 12, 2825-2830.

Prakash, K. B., Ruwali, A., & Kanagachidambaresan, G. (2021). *Introduction to Tensorflow Package. Programming with TensorFlow*. Springer. DOI: 10.1007/978-3-030-57077-4

Rampasek, L., & Goldenberg, A. (2016). TensorFlow: Biology's Gateway to Deep Learning? *Cell Systems*, *2*(1), 12–14. DOI: 10.1016/j.cels.2016.01.009 PMID: 27136685

Riddle, W. R., & Pickens, D. R. (2005). Extracting data from a DICOM file. *Medical Physics*, *32*(6Part1), 1537–1541. DOI: 10.1118/1.1916183 PMID: 16013711

Uludag, K., Wang, D. M., Mohamoud, Y., Wu, H. E., & Zhang, X. (2023). Antipsychotic-based machine learning models may help prediction of tardive dyskinesia in patients with schizophrenia. *Schizophrenia Research*, *252*, 33–35. DOI: 10.1016/j.schres.2022.12.026 PMID: 36621324

Virtanen, P., Gommers, R., Oliphant, T. E., Haberland, M., Reddy, T., Cournapeau, D., Burovski, E., Peterson, P., Weckesser, W., Bright, J., van der Walt, S. J., Brett, M., Wilson, J., Millman, K. J., Mayorov, N., Nelson, A. R. J., Jones, E., Kern, R., Larson, E., & Vázquez-Baeza, Y. (2020). SciPy 1.0: Fundamental algorithms for scientific computing in Python. *Nature Methods*, *17*(3), 261–272. DOI: 10.1038/s41592-019-0686-2 PMID: 32015543

Weber, R., Mangus, J. M., & Huskey, R. (2015). Brain Imaging in Communication Research: A Practical Guide to Understanding and Evaluating fMRI Studies. *Communication Methods and Measures*, *9*(1-2), 5–29. DOI: 10.1080/19312458.2014.999754

Yarberry, W. (2021). *Regular Expressions: Introduction. CRAN Recipes*. Springer.

# Chapter 8
# Identification of Hepatocellular Carcinoma From CT Liver Image for CAD Systems

**R. Prabakaran**
https://orcid.org/0000-0001-7804-4191
*MIT Campus, Anna University, India*

**D. Shiloah Elizabeth**
https://orcid.org/0000-0003-0543-584X
*CEG, Anna University, India*

**C. Sunil Retmin Raj**
https://orcid.org/0000-0002-5649-5466
*MIT Campus, Anna University, India*

## ABSTRACT

*Liver cancer is the sixth most common cancer. Liver cancer is frequently analyzed with Computed Tomography (CT) scan. Early diagnosis by CT could lead to high recovery rate, however going through all the CT slices for thousands or even millions of patients manually by professionals is hard, expensive, inefficient and prone to errors. The proposed method mainly focuses on precision of liver lesion segmentation and to improve performance of the CAD systems. The goal is to create a reliable Res-UNet model that can segment the liver, identify Regions of Interest (RoI) from nearby organs, and use that information to identify liver lesions using a different Res-UNet. The contributions of this chapter are identifying whether liver is healthy or affected by the disease (hepatocellular carcinoma), segmenting lesions from the*

DOI: 10.4018/979-8-3693-3840-7.ch008

*extracted liver image using Res-UNet and to evaluate the accuracy of segmentation. The developed models have been tested using LiTS dataset. The accuracy of liver segmentation and lesion segmentation achieved using ResUNet are 0.9918 and 0.99, respectively.*

## 1. INTRODUCTION

According to World Cancer Research Fund International (2020), liver cancer is the sixth most common cancer worldwide. There are two main forms of liver cancer: primary and secondary. Hepatocellular carcinoma (HCC) makes up 80% of cases of primary liver malignancies. Primary liver malignancies can be caused by hepatitis B and C viruses, obesity-related fatty liver disease, and alcohol-induced cirrhosis. Computed tomography (CT) scans can be used for diagnosis, and the ResUNet method, a combination of U-Net and ResNet model, can be used to segment the liver. The same technique can be used to isolate the lesions after the liver has been segmented. This method improves the ability to distinguish between the liver and the lesions, which helps doctors identify and diagnose liver lesions and create treatment regimens. For patients, it rises patient's chance of getting better treatment and decreases risk of lives owed to liver cancer.

### 1.1 Challenges

Usually, manual or semi-manual techniques are used to interpret liver CT scans; these approaches are labor-intensive, expensive, ineffective, and prone to a great deal of inaccuracy. In order to tackle these obstacles and enhance the accuracy of liver lesion identification, multiple computer-aided techniques have been created. However, due to a number of issues, including low contrast between the liver and surrounding organs, different contrast levels within lesions, varying lesion sizes and numbers, tissue abnormalities, and inconsistent lesion progression in response to treatment, these methods frequently have trouble segmenting the liver and lesions. These are formidable obstacles that must be surmounted.

### 1.2 Objective

The main objective is to use the ResUNet approach to improve the accuracy of liver and lesion segmentation. To capitalize on the advantages of both techniques, ResUNet combines the conventional U-Net and ResNet models by substituting residual blocks for convolution blocks. CNN training is made easier by the residual unit, and information transmission is prevented from degrading due to skip connections between the high and low levels of the network and within the residual unit. With this method, a neural network can be designed with a lot less parameters.

## 1.3 Scope of the Chapter

The main objectives of the suggested approach are to improve liver lesion segmentation accuracy and maximize the functionality of computer-aided diagnostic (CAD) systems. In a wide range of applications, including as image classification, object detection, face recognition, obstacle avoidance, material inspection, natural language processing, and more, convolutional neural networks (CNNs) have proven to produce remarkable results. The objective of this chapter is to present a reliable CNN model that can distinguish the region of interest (RoI) from neighbouring organs and segment the liver. Next, we'll utilize a second CNN and the ROI to find liver lesions.

## 1.4 Contribution

The contribution of work in this chapter are:

- To identify whether liver is healthy or affected by hepatocellular carcinoma (HCC).
- To segment lesions from the extracted liver image using Res-UNet
- To classify the affected cells using neural networks and to progress the accuracy of segmentation and performance of CAD systems.

## 1.5 Computed Tomography (CT)

Computed tomography (CT) emerged in the 1970s, revolutionizing diagnostic methods. It enhanced the accuracy of cancer diagnosis and treatment, improved surgical procedures, facilitated cardiac condition monitoring, and advanced injury treatment. CT scanning eliminated the necessity for exploratory surgeries. Unlike traditional x-ray machines with a static x-ray tube, CT scanners utilize a motorized x-ray source that rotates around the gantry. As the patient lies on a moving bed, the x-ray tube circles around them. CT scanners employ digital x-ray detectors instead of film, transmitting data to a computer. The computer processes this data from one rotation to construct a 2D image slice of the patient. It can also create a 3D model, allowing medical professionals to locate problems precisely by rotating the model or viewing successive slices.

## 1.6 HCC

HCC is one of the types of primary liver cancer. It occurs in people who have chronic liver ailments, like cirrhosis caused by hepatitis B or hepatitis C viruses. The danger of hepatocellular carcinoma is advanced in people with long-term liver ailments. It is also more common in people who drink huge amounts of alcohol and who have a collection of fat in the liver. Image testing such as CT and MRI is one of the methods to diagnose this disease.

## 1.7 CAD Systems

Medical image analysis is made easier for clinicians by computer-aided diagnostic (CAD) technologies. X-ray, MRI, and ultrasound diagnostic methods provide large amounts of data that radiologists and other medical professionals need to quickly evaluate. With the help of CAD systems, which analyze digital photos, experts may make better decisions by highlighting things like possible anomalies and identifying common patterns. Future uses for computer-aided diagnosis (CAD) in digital pathology are bright, especially with the advent of whole-slide imaging and machine learning methods.

## 1.8 CNN

CNNs are like neural networks as they are composed of neurons, where each neuron take several inputs take a weighted sum over them, then feeds the sum to the activation function where it responds with the output. They also have loss functions just like neural networks. Unlike neural networks CNNs operate on volumes rather than vectors. They also contain convolution layers which are basically of a set of independent filters that convolve on the input.

# 2. LITERATURE SURVEY

## 2.1 Segmentation of Liver Lesion for CAD

Lesion segmentation gives an achievement to assurance of the specific extent for CAD (Anwar et al., 2018). It is the huge advance in liver disease finding, action arranging and action assessment. LiTS - Liver Lesion Segmentation Challenge suggestions a normal test bed for looking at changed computerized liver injury segmentation methods. The primary utilization of image processing in digital is to build nature of images for conversion of human and machine understanding.

Lesion segmentation in liver CT volumes is referred as difficult errand as a result of the differing lesion surface & shape. Area of lesion represents test. Physical lesion segmentation is a slow process could be incorrect forever (Li et al., 2015). The examination aims to introduce an automated method for detecting lesions in each slice of volumetric computed tomography (CT) liver images. The proposed algorithm is fully automated for segmenting lesions, utilizing data from a series of CT volumes. The approach was tested on the LiTS 2017 dataset, employing metrics such as dice score, volumetric overlap error, and relative volume difference. An average dice score of 0.60 was achieved when evaluated on eighty CT volumes from the LiTS dataset, taking approximately 34 seconds per segment. Outcomes were huge, contrasted and cutting edge segmentation methods utilized in Computer aided design systems (Lee et al., 2007).

## 2.2 Low Dose Computed Tomography (LDCT) With a Residual Encoder-Decoder CNN

Given the expected danger of radiation of X-ray to patient, LDCT pulled in an impressive interest in the clinical imaging field (Chen et al., 2017). As of now, the fundamental LDCT methods incorporate merchant explicit sinogram space filtration & reiterative reconstruction algos, but must get raw data where designs aren't clear to more clients. Because of trouble of modelling arithmetic characteristic in image domain, current strategies to prepare for recreated images can't take out noise of image overall by keeping structural details. Inspired by possibility of deep learning, here they consolidate deconvolution organization, autoencoder & alternate route associations into remaining RED-CNN for LDCT imaging (Song et al., 2014). Later patchbased preparing, proposed RED-CNN accomplishes serious exhibition comparative with state of methods in stimulated and clinical cases. Particularly, this method was well assessed regarding suppression of noise, preservation of structures, and discovery of lesion (Xu et al., 2019).

## 2.3 Instinctive Segmentation of Liver Algorithm for CT Images

It is completely CNN in view of U-net projected to segment liver in CT scan images (Zhang & Xu, 2018). 2 amendments are made to first U-net building. Initially, additional pathway is added to unique net building to extract universal features & point features independently. Furthermore, quantity of convolutional channels of unique compression pathway, unique expansion pathway & new pathway is decreased (Dong et al., 2015). The 2 alterations made preparation faster & progress proficiency of convolution kernel feature extraction. At that point, results of segmentation previously & later alteration is thought about as far as execution, counting recollection rate & accuracy rate, to guarantee modified organization can go much complex than first net accuracy. From this point forward, it investigates details why the network can keep up good segmentation impact & sums up submission view of altered network (Chlebus et al., 2017).

## 2.4 LDCT Image Denoising Using GAN

This application incorporates recent advancements in deep learning methods, aiming to minimize mean-squared error (MSE) between denoised computed tomography (CT) images and their ground truth counterparts (Yang et al., 2018). Despite improvements in peak signal-to-noise ratio, methods based on weighted-MSE or MSE can compromise the visibility of key structural details after aggressive denoising. The application introduces a new CT image denoising approach based on GAN with Wasserstein distance and perceptual similarity (Christ et al., 2016). Wasserstein distance, a core concept in optimal transport theory, promises to enhance GAN performance. Perceptual loss mitigates noise by comparing the perceptual features of denoised output to those of the ground truth within an established feature space, while GAN additionally refines data distribution noise from solid to weaker statistical patterns. In this manner, our proposed technique moves our data on visual observation to denoising of image chore and, fit for diminishing the image noise level as well as likewise attempting to keep the critical information simultaneously. Promising outcomes have been acquired in tests with clinical CT images (Christ et al., 2016).

## 2.5 Segmentation of Liver in CT Built on ResUNet

This work suggests segmentation of liver basis utilizing ResUNet with 3D probabilistic & mathematical post process. The semantic segmentation model ResUNet enhances residual unit & batch normalization layer to up sampling & down testing some portion of U-Net for developing profounder network (Christ et al., 2016). For speedy converge, they proposed new loss function DCE, which is linearly shared by Dice misfortune & cross entropy misfortune. They used few CT images for training & testing to investigate more framework info. In the starting of ResUNet segmentation, completely associated 3Dimensional conditional random field was utilized to improve results of segmentation by discovering 2Dimensional neighbour regions & 3Dimensional volume info (Zhang et al., 2010). Finally, 3Dimensional connected mechanisms examining was used to stay some enormous parts & decrease noise of segmentation. Experimental results on LiTS shows that proposed system accomplish state of art presentation for segmentation of liver (Vancea et al., 2019).

## 2.6 Summary of Literature Survey

Earlier approaches composed of various CNN based models to segment liver and lesions from CT scan images. Models like UNet, RED-CNN, GAN networks, etc. are used in the previous works. These experiments uses various datasets like Chaos, MICCAI, LiTS, Sliver07, 3Dircadb datasets for segmentation. A ResUNet model has been proposed here, which is an advanced model to segment liver and lesion from CT scan images. ResUNet is a hybrid between the U-Net and ResNet models where it uses Residual impedes as opposed to customary convolution blocks. We utilized 2 Cascaded CNNs one for portioning the liver and extricating the ROI and the subsequent one utilize the extracted ROI from the first CNN and section the lesions. Though the researches done before failed to acquire clear image and provided image with high noise, the proposed method is used to attempt to bring highly defined image. It also improves evaluation metrics like dice score and accuracy.

# 3. METHODOLOGIES

## 3.1 System Architecture

ResUNet is used for liver segmentation and lesion segmentation, a semantic segmentation neural network which syndicates strengths of both UNet & ResNet. The combination of both UNet and residual neural network has two aids:

1) residual unit will comfort training of network
2) skip connections within a residual unit and among high levels and low levels of network will enable info broadcast without deprivation.

The network contains 3 parts:

- Encoding
- Bridge
- Decoding.

Encoding portion translates input image to solid illustrations. Decoding portion recovers representations to pixel classification. Bridge serves like a bridge linking decoding & encoding pathways. Encoding, bridge and decoding portions were made with residual units that consists of 2- 3 x 3 convolution blocks & identity mapping. Every convolution block embraces Batch Normalization layer, ReLU activation layer & convolutional layer. Identity mapping joins input & output of unit.

Encoding pathway has three residual units. In each unit, rather than using pooling process to down sample the feature map size, a stride of two is functional to the first convolution block to diminish the feature map by half.

Consistently, decoding pathway comprises of 3 residual units, too. Before each unit, there is up sampling of feature maps from inferior level & chain with the feature maps from the equivalent encoding pathway. Next last level of decoding pathway, 1x1 convolution & sigmoid activation layer used here to scheme multi-channel feature maps into chosen segmentation.

*Figure 1. ResUNet architecture*

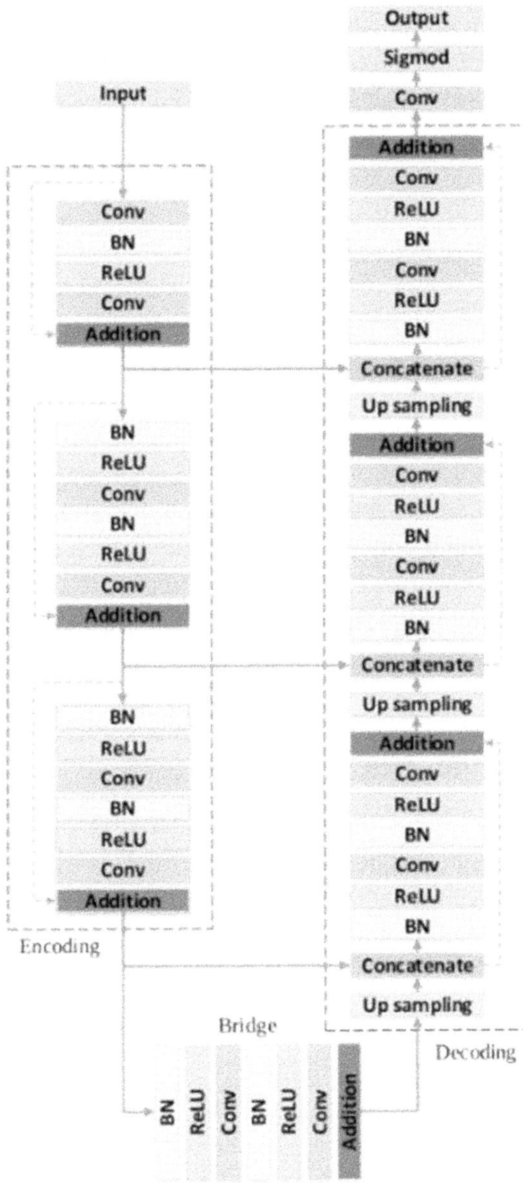

## 3.2 U-Net

The architecture contains of 2 main portions — left part is known as contracting pathway that is established by overall convolutional process; right part is known as expansive pathway that is established by transposed 2dimensional convolutional layers.

### Contracting Path

Every process establishes 2 convolutional layers, max pooling method which splits the dimensions of image. The method is repeated 3 more times. And then 2 convolutional layers are made, with no max pooling

### Expansive Path

An image is enlarged to its original proportions throughout this process. An upsampling method called transposed convolution makes images larger. The original image is padded, and a convolution technique is then applied. Following the transposed convolution, the image's size rises to 56x56x512 from 28x28x1024, and it is concatenated with an image from the contracting path to produce an image that is 56x56x1024. A more accurate prediction can be made by combining data from preceding layers. This procedure is carried out several times. The image must be modified in the last stage to satisfy the computation specifications. A convolution layer with a single, 1x1 filter makes up the layer before it. The remaining steps are the same as for regular neural network training.

*Figure 2. UNet architecture*

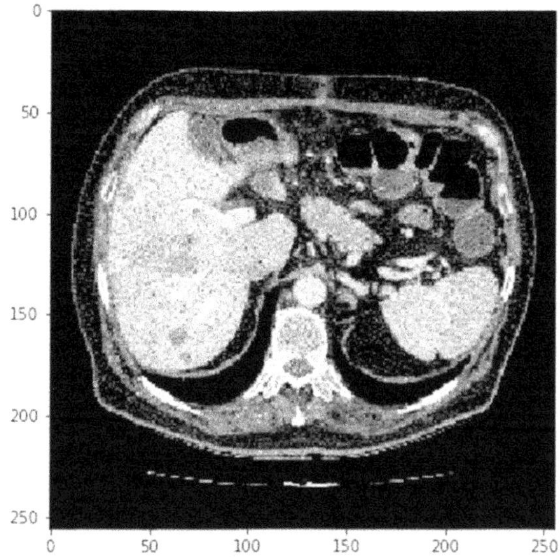


*Figure 3. Basic neural element used in U-Net*

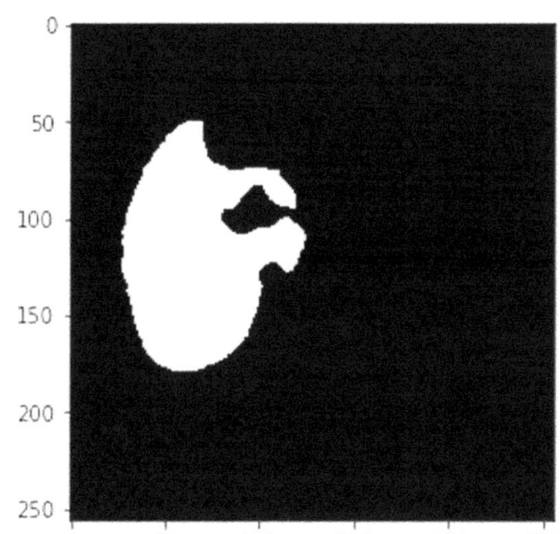

*Figure 4. Residual element in proposed ResUNet.*

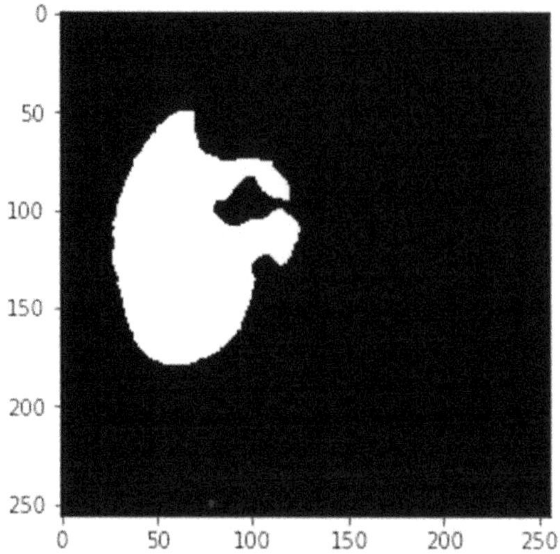


## 3.3 Residual Neural Network

A specific neural network called Residual Neural Network (ResNet) was unveiled by Kaiming He, Xiangyu Zhang, Shaoqing Ren, and Jian Sun in 2015. ResNet uses residual blocks, which include skip connections, to overcome the difficulty of training very deep networks. These skip connections give data a new route through the network by avoiding specific layers—which differ between models. This bypass modifies the layer's output and is essential to the residual blocks. The input 'x' would just be multiplied by the weights of the layer and summed with a bias term in the absence of these skip connections. ResNet's skip connections facilitate easier gradient flow, which helps address the issue of vanishing gradients in deep neural networks. Furthermore, the model may learn identity functions thanks to these connections, guaranteeing that higher levels operate just as well as or better than lower layers. ResNet makes use of a 34-layer plain network design that was modified to become a residual network by adding shortcut connections. This architecture was influenced by the VGG-19. Figure 5 provides an illustration of this conversion.

*Figure 5. ResNet architecture*

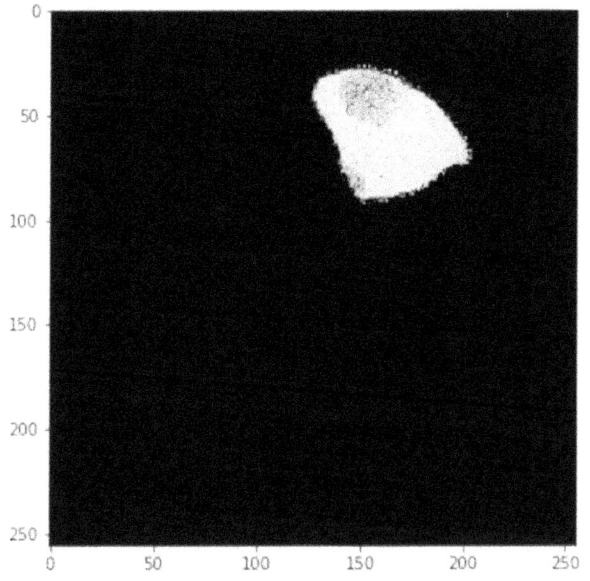


# 4. ALGORITHM DEVELOPMENT AND IMPLEMENTATION

## 4.1 Data Pre-Processing

Pre-processing is needed to every slice of the input image so as to differentiate the liver from neighbouring Organs.

### Hounsfield Windowing

The average attenuation values of a given voxel are measured using the Hounsfield Unit (HU), which is based on the attenuation values of distilled water at normal temperature and pressure (water has a HU of zero and air has a HU of -1000). Twelve bits are used to encode HU, yielding 4,096 values between -1024 and 3071 HU. The formula to compute the HU of any tissue is

$$HU=1000x(\mu\_tissue-\mu\_H2O)/\mu\_H2O$$

where the linear attenuation coefficient is denoted by μ. Higher values correspond to brighter pixels. Each pixel's HU value is represented by a grayscale intensity. Values in the range [-100, 400] are subjected to Hounsfield windowing once the slides are read in DICOM format.

## 4.2 Existing System Approach

CNNs are equivalent to neural networks as they were made up of neurons, where each neuron take several inputs take a weighted sum over them, then feeds the sum to the activation function where it responds with the output. They also have loss functions just like neural networks. Unlike neural networks CNNs operate on volumes rather than vectors. They also contain convolution layers which are basically of a set of independent filters that convolve on the input image.

*Table 1. HU values for body organs*

| | **HU values for body organs** |
|---|---|
| Air | -1000 |
| Blood | 40 |
| Bones<br>Cerebrospinal | 1000<br>15 |
| Fat | -50 to -100 |
| Grey Matter | 43 |
| Kidney<br>Liver | 30<br>40 to 60 |
| Muscle | 10 to 40 |
| Water<br>White Matter | 0<br>46 |

### 4.2.1 U-Net Algorithm

In the existing system, the liver segmentation algorithm is based on CNN models like UNet architecture.

The UNet architecture comprises of two main parts:

1. Contracting path, which is created by the general convolutional process;
2. Expansive path, which is created by transposed 2d convolutional layers.

## Contracting Path

The contracting path, sometimes referred to as the encoder or down-sampling path, is made up of the following four blocks:

conv_layer1 -> conv_layer2 -> max_pooling -> dropout (optional)

Every block consists of:

- A 3x3 Convolution Layer with the ReLU function on.
- An additional 3 x 3 Convolution Layer including the ReLU activation mechanism.
- a Max Pooling Layer of 2x2.
- A Dropout Layer (0.5 dropout rate) is an optional feature.

## Expanding Path

The expanding path is made up of four blocks and is sometimes referred to as the up-sampling path or decoder. Every block consists of:

- a layer of deconvolution with two steps.
- Connecting with the matching clipped feature map from the path of contraction.
- an activation function of ReLU in a 3x3 convolution layer.
- An additional 3x3 convolution layer with an activation function of ReLU.

*Figure 6. UNet algorithm*

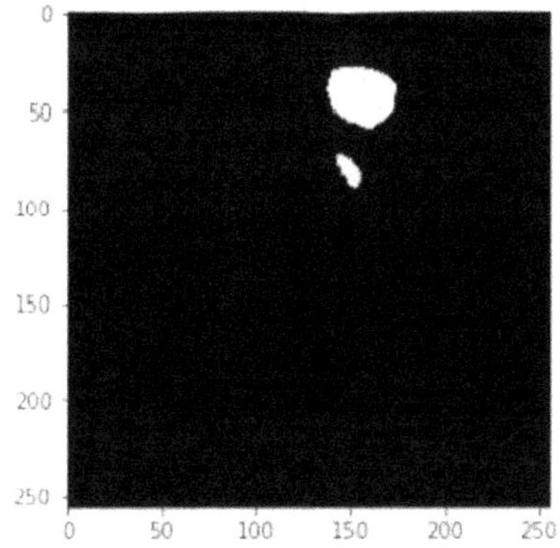

## 4.3 Proposed System Approach

ResUNet is the foundation of the Liver and Lesion Segmentation algorithm. ResUNet is a cross between the classic U-Net and ResNet models, using residual blocks in place of convolution blocks.

The primary benefit of ResUNet is that it will improve CNN training, skip connections between high- and low-level network nodes, and remove deterioration during information propagation, resulting in fewer training parameters inside the residual unit.

1. 2- 3x3 convolution blocks that make up the residual portion are composed of the following: the batch normalization layer, the ReLU activation layer, and the convolutional layer.
2. Charting of Identity

This extending path uses up-convolution layers to add spatial information to the segmentation image and regain the feature map size. The course relative info from contracting pathway will be shifted to up sampling pathway by skipping connections.

*Figure 7. Different blocks of algorithm*

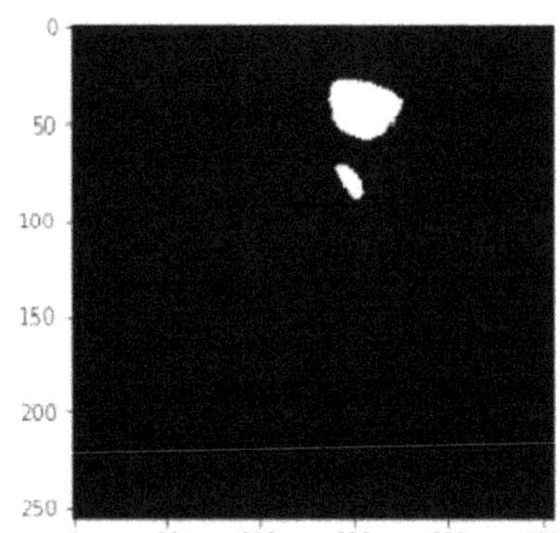

*Figure 8. ResUNet algorithm*

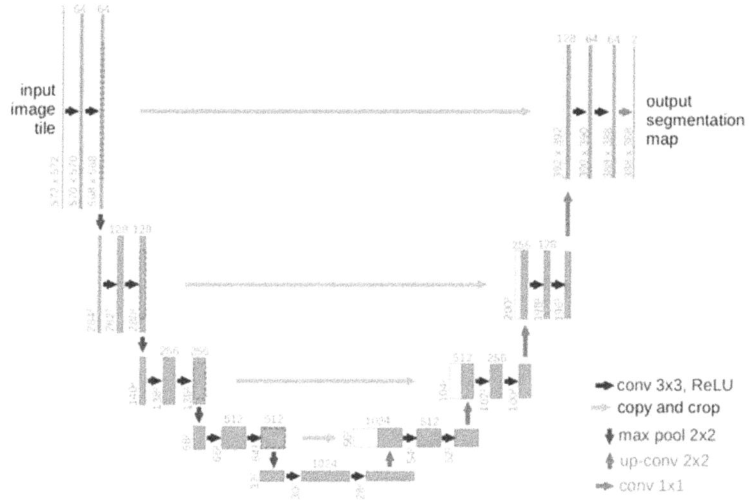


## 4.4 Training

The unique Res-UNet architecture is used in the construction of the network. The segmentation masks that corresponded to the input photos were used to train the model. For training, a total of 2,346 pictures with data augmentation were used from two datasets. The network's hyperparameters, including the learning rate, batch size, number of epochs, filter count, validation split, dropout rate, optimizer, loss function, and activation function, were all varied during training and tested at different values.

## 4.5 Optimizer

The Adam Optimizer provides flexibility for both RMSprops (root mean squared) and stochastic gradient descent (SGD). It is a method for efficient stochastic optimization that requires little memory and only first-order gradients. It determines distinct adaptive learning rates for every network parameter. An Adam optimizer with a learning rate of 0.0001 was employed in this work.

## 4.6 Dataset

The publicly available 3D-IRCADb01 dataset, provided by the IRCAD Research Institute against digestive cancer, includes 3D CT scans of ten men and ten women. The CT scans are in DICOM format and are divided into 2D slices, amounting to approximately 2,800 slices in total. Each slice is accompanied by masks for the liver, lesions, bones, arteries, kidneys, and lungs.

# 5. RESULTS AND DISCUSSION

## 5.1 Implementation Environment

Many Python programs and libraries are utilized for lesion and liver segmentation, such as:

1. Anaconda, which simplifies deployment and package management.
2. NumPy is a Python scientific computing toolkit that supports multidimensional matrices and arrays and offers mathematical functions for manipulating these kinds of objects.
3. Matplotlib is a library used for graphing and data visualization.
4. The Google Brain Team created TensorFlow, an open-source library for deep neural network and machine learning research.
5. Keras is a machine learning and deep learning tool that makes complex tasks easier for non-experts. It may be used with a variety of backend engines, including TensorFlow, CNTK, and Theano, without requiring a thorough understanding of their internal structures or syntax.

## 5.2 Evaluation Metrics

To assess the performance of the model, the following metrics are used:

1. Dice Coefficient: The overlap between the predicted labels B and the input classes A is measured by this statistic. The following formula is used to compute it:

$$Dice = (2 \times |A \cap B|) / (|A| + |B|)$$

2. True Value Accuracy: This measure evaluates the precision of segmenting liver and lesions and is computed as follows:

True Value Accuracy = (True Positives / (True Positives + False Negatives)) × 100%

## 5.3 Liver Segmentation Results

*Figure 9. Performance evaluation – Dice Coef, Val Dice Coef, Loss and Val*

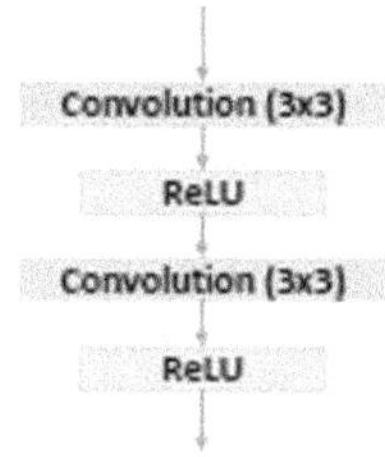


Random sample has been analysed using the model, as shown in figure 10. Figure 11 illustrates the liver's real label, and Figure 12 shows the result.

*Figure 10. CT scan of a random sample before liver segmentation*

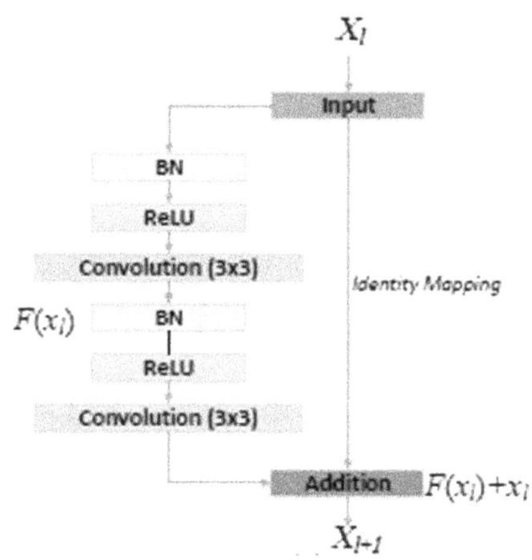

*Figure 11. True label of the liver*

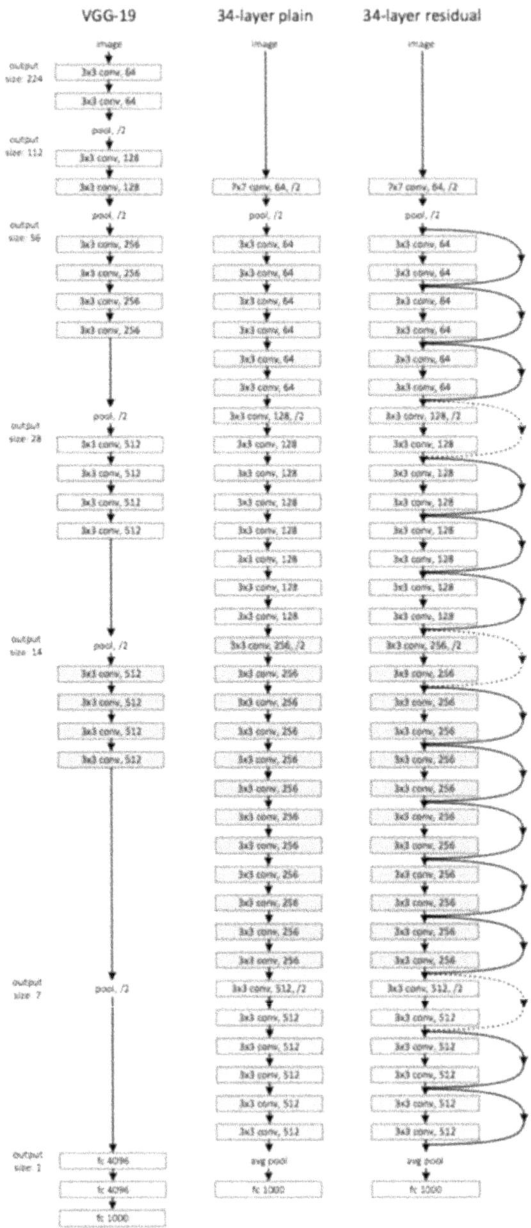

*Figure 12. Predicted label of the liver*

```
def unet(img_channels, image_rows, image_cols, neurons=16):
    inputs = Input((img_channels, image_rows, image_cols))
    conv1 = Conv2D(neurons*1, (3, 3), activation='relu', padding='same')(inputs)
    conv1 = Conv2D(neurons*1, (3, 3), activation='relu', padding='same')(conv1)
    conv1 = BatchNormalization()(conv1)
    pool1 = MaxPooling2D(pool_size=(2, 2))(conv1)
    pool1 = Dropout(0.25)(pool1)

    conv2 = Conv2D(neurons*2, (3, 3), activation='relu', padding='same')(pool1)
    conv2 = Conv2D(neurons*2, (3, 3), activation='relu', padding='same')(conv2)
    conv2 = BatchNormalization()(conv2)
    pool2 = MaxPooling2D(pool_size=(2, 2))(conv2)
    pool2 = Dropout(0.5)(pool2)

    conv3 = Conv2D(neurons*4, (3, 3), activation='relu', padding='same')(pool2)
    conv3 = Conv2D(neurons*4, (3, 3), activation='relu', padding='same')(conv3)
    conv3 = BatchNormalization()(conv3)
    pool3 = MaxPooling2D(pool_size=(2, 2))(conv3)
    pool3 = Dropout(0.5)(pool3)

    conv4 = Conv2D(neurons*8, (3, 3), activation='relu', padding='same')(pool3)
    conv4 = Conv2D(neurons*8, (3, 3), activation='relu', padding='same')(conv4)
    conv4 = BatchNormalization()(conv4)
    pool4 = MaxPooling2D(pool_size=(2, 2))(conv4)
    pool4 = Dropout(0.5)(pool4)

    conv5 = Conv2D(neurons*16, (3, 3), activation='relu', padding='same')(pool4)
    conv5 = Conv2D(neurons*16, (3, 3), activation='relu', padding='same')(conv5)

    up6 = concatenate([UpSampling2D(size=(2, 2))(conv5), conv4], axis=1)
    up6 = Dropout(0.5)(up6)
    conv6 = Conv2D(neurons*8, (3, 3), activation='relu', padding='same')(up6)
    conv6 = Conv2D(neurons*8, (3, 3), activation='relu', padding='same')(conv6)

    up7 = concatenate([UpSampling2D(size=(2, 2))(conv6), conv3], axis=1)
    up7 = Dropout(0.5)(up7)
    conv7 = Conv2D(neurons*4, (3, 3), activation='relu', padding='same')(up7)
    conv7 = Conv2D(neurons*4, (3, 3), activation='relu', padding='same')(conv7)

    up8 = concatenate([UpSampling2D(size=(2, 2))(conv7), conv2], axis=1)
    up8 = Dropout(0.5)(up8)
    conv8 = Conv2D(neurons*2, (3, 3), activation='relu', padding='same')(up8)
    conv8 = Conv2D(neurons*2, (3, 3), activation='relu', padding='same')(conv8)

    up9 = concatenate([UpSampling2D(size=(2, 2))(conv8), conv1], axis=1)
    up9 = Dropout(0.5)(up9)
    conv9 = Conv2D(neurons*1, (3, 3), activation='relu', padding='same')(up9)
    conv9 = Conv2D(neurons*1, (3, 3), activation='relu', padding='same')(conv9)

    conv10 = Dropout(0.5)(conv9)
    conv10 = Conv2D(1, (1, 1), activation='sigmoid')(conv10)
```

## 5.4 Lesion Segmentation Results

Using the liver mask shown in figure 13, we randomly selected a sample and ran it through the model. Figure 14 shows the actual label of the lesion, and Figure 15 shows the outcome.

*Figure 13. CT scan with the liver mask applied upon it before lesion segmentation*

```
def conv_block(x, filters, kernel_size=(3, 3), padding="same", strides=1):
    conv = bn_act(x)
    conv = keras.layers.Conv2D(filters, kernel_size, padding=padding, strides=strides)(conv)
    return conv

def stem(x, filters, kernel_size=(3, 3), padding="same", strides=1):
    conv = keras.layers.Conv2D(filters, kernel_size, padding=padding, strides=strides)(x)
    conv = conv_block(conv, filters, kernel_size=kernel_size, padding=padding, strides=strides)

    shortcut = keras.layers.Conv2D(filters, kernel_size=(1, 1), padding=padding, strides=strides)(x)
    shortcut = bn_act(shortcut, act=False)

    output = keras.layers.Add()([conv, shortcut])
    return output

def residual_block(x, filters, kernel_size=(3, 3), padding="same", strides=1):
    res = conv_block(x, filters, kernel_size=kernel_size, padding=padding, strides=strides)
    res = conv_block(res, filters, kernel_size=kernel_size, padding=padding, strides=1)

    shortcut = keras.layers.Conv2D(filters, kernel_size=(1, 1), padding=padding, strides=strides)(x)
    shortcut = bn_act(shortcut, act=False)

    output = keras.layers.Add()([shortcut, res])
    return output

def upsample_concat_block(x, xskip):
    u = keras.layers.UpSampling2D((2, 2))(x)
    c = keras.layers.Concatenate()([u, xskip])
    return c
```

*Figure 14. True label of the lesion*

```
def ResUNet():
    f = [16, 32, 64, 128, 256]
    inputs = keras.layers.Input((image_size, image_size, 3))

    ## Encoder
    e0 = inputs
    e1 = stem(e0, f[0])
    e2 = residual_block(e1, f[1], strides=2)
    e3 = residual_block(e2, f[2], strides=2)
    e4 = residual_block(e3, f[3], strides=2)
    e5 = residual_block(e4, f[4], strides=2)

    ## Bridge
    b0 = conv_block(e5, f[4], strides=1)
    b1 = conv_block(b0, f[4], strides=1)

    ## Decoder
    u1 = upsample_concat_block(b1, e4)
    d1 = residual_block(u1, f[4])

    u2 = upsample_concat_block(d1, e3)
    d2 = residual_block(u2, f[3])

    u3 = upsample_concat_block(d2, e2)
    d3 = residual_block(u3, f[2])

    u4 = upsample_concat_block(d3, e1)
    d4 = residual_block(u4, f[1])

    outputs = keras.layers.Conv2D(1, (1, 1), padding="same", activation="sigmoid")(d4)
    model = keras.models.Model(inputs, outputs)
    return model
```

*Figure 15. Predicted label of the lesion*

```
Epoch 11/20
2027/2027 [==============================]2027/2027 [==============================] - 733s 362ms/step - loss: 0.0394 - acc: 0.9862 - dic
e_coef: 0.9606 - val_loss: 0.1244 - val_acc: 0.9721 - val_dice_coef: 0.8756

Epoch 12/20
2027/2027 [==============================]2027/2027 [==============================] - 732s 361ms/step - loss: 0.0359 - acc: 0.9866 - dic
e_coef: 0.9641 - val_loss: 0.0668 - val_acc: 0.9818 - val_dice_coef: 0.9332

Epoch 13/20
2027/2027 [==============================]2027/2027 [==============================] - 732s 361ms/step - loss: 0.0305 - acc: 0.9871 - dic
e_coef: 0.9695 - val_loss: 0.0412 - val_acc: 0.9853 - val_dice_coef: 0.9588

Epoch 14/20
2027/2027 [==============================]2027/2027 [==============================] - 732s 361ms/step - loss: 0.0313 - acc: 0.9871 - dic
e_coef: 0.9687 - val_loss: 0.0451 - val_acc: 0.9848 - val_dice_coef: 0.9549

Epoch 15/20
2027/2027 [==============================]2027/2027 [==============================] - 734s 362ms/step - loss: 0.0292 - acc: 0.9873 - dic
e_coef: 0.9708 - val_loss: 0.2057 - val_acc: 0.9647 - val_dice_coef: 0.7943

Epoch 16/20
2027/2027 [==============================]2027/2027 [==============================] - 734s 362ms/step - loss: 0.0277 - acc: 0.9875 - dic
e_coef: 0.9723 - val_loss: 0.0263 - val_acc: 0.9873 - val_dice_coef: 0.9737

Epoch 17/20
2027/2027 [==============================]2027/2027 [==============================] - 732s 361ms/step - loss: 0.0282 - acc: 0.9874 - dic
e_coef: 0.9718 - val_loss: 0.0264 - val_acc: 0.9873 - val_dice_coef: 0.9736

Epoch 18/20
2027/2027 [==============================]2027/2027 [==============================] - 732s 361ms/step - loss: 0.0246 - acc: 0.9879 - dic
e_coef: 0.9754 - val_loss: 0.0248 - val_acc: 0.9875 - val_dice_coef: 0.9752

Epoch 19/20
2027/2027 [==============================]2027/2027 [==============================] - 732s 361ms/step - loss: 0.0240 - acc: 0.9879 - dic
e_coef: 0.9760 - val_loss: 0.0268 - val_acc: 0.9873 - val_dice_coef: 0.9732

Epoch 20/20
2027/2027 [==============================]2027/2027 [==============================] - 733s 362ms/step - loss: 0.0242 - acc: 0.9879 - dic
e_coef: 0.9758 - val_loss: 0.0224 - val_acc: 0.9878 - val_dice_coef: 0.9776
```

## 5.5 Comparative Analysis of Liver and Lesion Segmentation

A comparative analysis has been made with previous works. The level of dice score for liver segmentation and lesion segmentation have been compared in table 2. It can be seen that there is an improvement of the dice score using the ResUNet method.

*Table 2. Comparison of liver segmentation*

| **Title/Year** | **Dice Score** | |
|---|---|---|
| | **Liver Segmentation** | **Lesion Segmentation** |
| Automatic liver lesion segmentation in CT with fully CNN & object-based post processing - 2018 | 0.96 | 0.68 |
| Automatic Liver & Lesion Segmentation in CT Using Cascaded Fully CNN & 3Dimensional Conditional Random Fields - 2016 | 0.94 | 0.82 |
| Cascaded deep convolutional encoder decoder neural networks for efficient liver tumor segmentation - 2019 | 0.95 | 0.64 |
| Proposed work – ResUNet | 0.97 | 0.87 |

The accuracy of liver segmentation and lesion segmentation achieved using ResUNet are 0.9918 and 0.99, respectively.

## 6. CONCLUSION AND FUTURE WORKS

In this work, computed tomography scan pictures were used to perform automatic segmentation of the liver and lesions using the ResUNet model. As can be seen from the results section, CNNs have been found to be the most effective techniques for segmenting liver lesions.

The work has been implemented for detection of lesions from 2D slices of CT scan images. It can be extended to 3D images and other imaging modalities as well. Moreover, in future the model can be built so as to classify the type of lesion and stages as well.

## REFERENCES


Ahmad, M., Ai, D., Xie, G., Qadri, S. F., Song, H., Huang, Y., Wang, Y., & Yang, J. (2019). Deep belief network modeling for automatic liver segmentation. *IEEE Access : Practical Innovations, Open Solutions*, *7*, 20585–20595. DOI: 10.1109/ACCESS.2019.2896961

Anwar, S. M., Awan, S., Yousaf, S., & Majid, M. (2018, December). Segmentation of liver tumor for computer aided diagnosis. In *2018 IEEE-EMBS Conference on Biomedical Engineering and Sciences (IECBES)* (pp. 366-370). IEEE. DOI: 10.1109/IECBES.2018.8626682

Budak, Ü., Guo, Y., Tanyildizi, E., & Şengür, A. (2020). Cascaded deep convolutional encoder-decoder neural networks for efficient liver tumor segmentation. *Medical Hypotheses*, *134*, 109431. DOI: 10.1016/j.mehy.2019.109431 PMID: 31669758

Chen, H., Zhang, Y., Kalra, M. K., Lin, F., Chen, Y., Liao, P., Zhou, J., & Wang, G. (2017). Low-dose CT with a residual encoder-decoder convolutional neural network. *IEEE Transactions on Medical Imaging*, *36*(12), 2524–2535. DOI: 10.1109/TMI.2017.2715284 PMID: 28622671

Chlebus, G., Meine, H., Moltz, J. H., & Schenk, A. (2017). Neural network-based automatic liver tumor segmentation with random forest-based candidate filtering. *arXiv preprint arXiv:1706.00842*.

Chlebus, G., Schenk, A., Moltz, J. H., van Ginneken, B., Hahn, H. K., & Meine, H. (2018). Automatic liver tumor segmentation in CT with fully convolutional neural networks and object-based postprocessing. *Scientific Reports*, *8*(1), 15497. DOI: 10.1038/s41598-018-33860-7 PMID: 30341319

Christ, P. F., Elshaer, M. E. A., Ettlinger, F., Tatavarty, S., Bickel, M., Bilic, P., & Menze, B. H. (2016). Automatic liver and lesion segmentation in CT using cascaded fully convolutional neural networks and 3D conditional random fields. In *Medical Image Computing and Computer-Assisted Intervention–MICCAI 2016: 19th International Conference, Athens, Greece, October 17-21, 2016* [Springer International Publishing.]. *Proceedings*, *19*(Part II), 415–423.

Dong, C., Loy, C. C., He, K., & Tang, X. (2015). Image super-resolution using deep convolutional networks. *IEEE Transactions on Pattern Analysis and Machine Intelligence*, *38*(2), 295–307. DOI: 10.1109/TPAMI.2015.2439281 PMID: 26761735

Erdt, M., Steger, S., Kirschner, M., & Wesarg, S. (2010, October). Fast automatic liver segmentation combining learned shape priors with observed shape deviation. In *2010 IEEE 23rd International Symposium on Computer-Based Medical Systems (CBMS)* (pp. 249-254). IEEE. DOI: 10.1109/CBMS.2010.6042650

Goryawala, M., Gulec, S., Bhatt, R., McGoron, A. J., & Adjouadi, M. (2014). A low-interaction automatic 3D liver segmentation method using computed tomography for selective internal radiation therapy. *BioMed Research International, 2014*, 2014. DOI: 10.1155/2014/198015 PMID: 25105118

Heimann, T., Van Ginneken, B., Styner, M. A., Arzhaeva, Y., Aurich, V., Bauer, C., Beck, A., Becker, C., Beichel, R., Bekes, G., Bello, F., Binnig, G., Bischof, H., Bornik, A., Cashman, P., Ying Chi, , Cordova, A., Dawant, B. M., Fidrich, M., & Wolf, I. (2009). Comparison and evaluation of methods for liver segmentation from CT datasets. *IEEE Transactions on Medical Imaging*, *28*(8), 1251–1265. DOI: 10.1109/TMI.2009.2013851 PMID: 19211338

Lamecker, H., Lange, T., & Seebass, M. (2004). Segmentation of the liver using a 3D statistical shape model.

Lee, J., Kim, N., Lee, H., Seo, J. B., Won, H. J., Shin, Y. M., & Shin, Y. G. (2007). Efficient liver segmentation exploiting level-set speed images with 2.5 D shape propagation. In *Proceedings of the MICCAI Workshop on 3-D Segmentat. Clinic: A Grand Challenge* (pp. 189-196).

Li, G., Chen, X., Shi, F., Zhu, W., Tian, J., & Xiang, D. (2015). Automatic liver segmentation based on shape constraints and deformable graph cut in CT images. *IEEE Transactions on Image Processing*, *24*(12), 5315–5329. DOI: 10.1109/TIP.2015.2481326 PMID: 26415173

Linguraru, M. G., Richbourg, W. J., Watt, J. M., Pamulapati, V., & Summers, R. M. (2012). Liver and tumor segmentation and analysis from CT of diseased patients via a generic affine invariant shape parameterization and graph cuts. In *Abdominal Imaging. Computational and Clinical Applications: Third International Workshop, Held in Conjunction with MICCAI 2011, Toronto, ON, Canada, September 18, 2011, Revised Selected Papers 3* (pp. 198-206). Springer Berlin Heidelberg. DOI: 10.1007/978-3-642-28557-8_25

Pan, S., & Dawant, B. M. (2001, July). Automatic 3D segmentation of the liver from abdominal CT images: a level-set approach. In *Medical Imaging 2001: Image Processing* (Vol. 4322, pp. 128-138). SPIE.

Rusko, L., Bekes, G., Nemeth, G., & Fidrich, M. (2007). Fully automatic liver segmentation for contrast-enhanced CT images. *MICCAI Wshp. 3D Segmentation in the Clinic: A Grand Challenge, 2*(7), 143-150.

Song, H., Zhang, Q., & Wang, S. (2014, November). Liver segmentation based on SKFCM and improved GrowCut for CT images. In *2014 IEEE international conference on bioinformatics and biomedicine (BIBM)* (pp. 331-334). IEEE. DOI: 10.1109/BIBM.2014.6999179

Vancea, F., Mitrea, D., Nedevschi, S., Rotaru, M., Stefanescu, H., & Badea, R. (2019, September). Hepatocellular carcinoma segmentation within ultrasound images using convolutional neural networks. In *2019 IEEE 15th International Conference on Intelligent Computer Communication and Processing (ICCP)* (pp. 483-490). IEEE. DOI: 10.1109/ICCP48234.2019.8959687

Xu, W., Liu, H., Wang, X., & Qian, Y. (2019, July). Liver segmentation in CT based on ResUNet with 3D probabilistic and geometric post process. In *2019 IEEE 4th International Conference on Signal and Image Processing (ICSIP)* (pp. 685-689). IEEE.

Yang, Q., Yan, P., Zhang, Y., Yu, H., Shi, Y., Mou, X., Kalra, M. K., Zhang, Y., Sun, L., & Wang, G. (2018). Low-dose CT image denoising using a generative adversarial network with Wasserstein distance and perceptual loss. *IEEE Transactions on Medical Imaging, 37*(6), 1348–1357. DOI: 10.1109/TMI.2018.2827462 PMID: 29870364

Zhang, G., Tzoumas, S., Cheng, K., Liu, F., Liu, J., Luo, J., Bai, J., & Xing, L. (2017). Generalized adaptive Gaussian Markov random field for x-ray luminescence computed tomography. *IEEE Transactions on Biomedical Engineering, 65*(9), 2130–2133. DOI: 10.1109/TBME.2017.2785364 PMID: 29989945

Zhang, L., & Xu, L. (2018, June). An automatic liver segmentation algorithm for CT images U-net with separated paths of feature extraction. In *2018 IEEE 3rd International Conference on Image, Vision and Computing (ICIVC)* (pp. 294-298). IEEE. DOI: 10.1109/ICIVC.2018.8492721

Zhang, X., Tian, J., Deng, K., Wu, Y., & Li, X. (2010). Automatic liver segmentation using a statistical shape model with optimal surface detection. *IEEE Transactions on Biomedical Engineering, 57*(10), 2622–2626. DOI: 10.1109/TBME.2010.2056369 PMID: 20615804

# Chapter 9
# Secure and Robust Digital Image Watermarking Technique to Secure Medical Images: A Comprehensive Approach

**Chandan Kumar**
https://orcid.org/0000-0001-9163-3206
*Career Point University, Hamirpur, India*

## ABSTRACT

*This paper presents a novel watermarking method that utilizes the combined power of the Discrete Wavelet Transform (DWT), Discrete Cosine Transform (DCT),and Set Partitioning in Hierarchical Trees (SPIHT) algorithm. Our method embeds two distinct watermarks,into different sub-bands of host image.Initially, the cover image and Watermark1 undergo optimization using hybrid optimization techniques(Particle Swarm Optimization and the Butterfly Optimization). Subsequently, Watermark1 is encrypted using the Arnold transform, while Watermark 2 is encoded using a Hamming encoder. Both encrypted watermarks are then embedded into their respective sub-bands.To compress the watermarked image,while retaining the embedded watermarks, we apply the SPIHT compression algorithm. The performance of our proposed method is evaluated against various attacks.Results demonstrate the robustness of the proposed method across these attacks.Notably, the proposed method achieves a maximum PSNR value of 48.50 dB and a maximum NC value of 0.9999, underscoring similarity between the original and watermarked images.*

DOI: 10.4018/979-8-3693-3840-7.ch009

## 1. INTRODUCTION

Digital watermarking techniques have become increasingly essential in safeguarding multimedia content, especially concerning intellectual property rights and copyright protection. In today's digital landscape, where content replication and unauthorized distribution are rampant, the need for effective protection mechanisms has grown exponentially. Digital watermarking provides a solution by embedding hidden identifiers within digital media, thus establishing ownership and deterring unauthorized usage. Recent years have seen a surge in the development of robust and secure watermarking methods capable of withstanding various attacks while preserving the integrity of the host image(Kumar,218).In this study, we proposed a resilient watermarking method that integrates the Discrete Wavelet Transform (DWT), Discrete Cosine Transform (DCT), and Set Partitioning in Hierarchical Trees (SPIHT) techniques. Our approach involves decomposing the host image into distinct sub-bands, where two secure watermarks are embedded. Initially, both the cover image and Watermark 1 undergo optimization using a Hybrid optimization technique, which combines the whale optimization and cuckoo search algorithms. Subsequently, Watermark 1 and Watermark 2 are encrypted using the Arnold Transform and Hamming Encoder, respectively, before being embedded into the cover image. Finally, SPIHT compression is applied to minimize distortion while maintaining the watermark's integrity(Singh,2023)

The proposed watermarking technique's performance is rigorously evaluated against a spectrum of watermarking attacks, including JPEG compression, Gaussian noise, and salt and pepper noise. Our findings demonstrate the method's robustness in preserving the embedded watermarks under various adversarial scenarios while ensuring minimal distortion to the cover image.

Key contributions of our research include:

1. Robust Watermarking Method: Our method exhibits high resilience against diverse watermarking attacks, owing to the synergistic combination of DWT, DCT, SPIHT techniques, and the hybrid optimization approach.
2. Minimal Distortion: Despite the watermark embedding process, our method maintains the visual quality and integrity of the cover image, ensuring a seamless viewing experience.
3. Enhanced Security Measures: Additional security measures, such as watermark encryption and encoding, fortify the embedded watermarks against unauthorized access and tampering, bolstering authenticity verification.
4. Novel Hybrid Optimization Technique: The introduction of a hybrid optimization technique streamlines the optimization process, leading to improved robustness and efficiency.

5. Comprehensive Evaluation: Through extensive experimentation and comparisons, we provide a thorough assessment of our method's performance, highlighting its superiority in terms of robustness and image quality preservation.

In summary, our research presents a novel watermarking methodology that addresses the challenges of digital multimedia security. By combining advanced techniques with enhanced security measures, our method offers a promising solution for protecting multimedia content in various applications.

## 2. LITERATURE SURVEY

In recent years, there has been a notable surge in research focusing on the development of watermarking systems tailored specifically for medical images. Saber et al.(2023) introduced a watermarking system leveraging optimization algorithms, presenting advancements that align with the growing trend of digital image watermarking practices. Their study underscores the increasing importance of protecting medical image data, particularly amidst the evolving landscape of digital technologies. Zeng et al.(2024) proposed a multi-watermarking algorithm based on KAZE-DCT, showcasing the significance of feature extraction and encryption techniques in medical image watermarking. By emphasizing watermark embedding and extraction processes, their approach contributes to enhancing the security and robustness of medical image data. Vaidya(2023) addressed the need for robust watermarking methods in medical imaging by introducing a fingerprint-based approach in a hybrid transform domain. This novel method offers enhanced security and resilience against unauthorized access, highlighting the potential of hybrid transform-based techniques in safeguarding medical image integrity. Furthermore, Pallaw et al. (2023) presented a watermarking scheme based on nature-inspired optimization for telemedicine applications, demonstrating the relevance of optimization algorithms in ensuring secure transmission and medical images authentication. Their study underscores the importance of nature-inspired optimization techniques in developing robust watermarking solutions for telemedicine scenarios. Sinhal and Ansari (2023) explored the application of machine learning in multipurpose medical image watermarking, showcasing the role of deep neural networks in providing effective security solutions. By leveraging machine learning algorithms, their framework offers versatility and adaptability in watermark embedding processes, contributing to improved data security in medical imaging applications. Additionally, Sheng et al.(2023) proposed a zero watermarking algorithm based on Resnet 50-DCT, presenting a novel approach to feature extraction and watermark embedding in medical images. Their study highlights the potential of deep learning techniques in developing innovative watermarking methods for medical image protection. Gull and Parah(2024) conducted a comprehensive review of advances in medical image watermarking, providing insights into the evolving landscape of watermarking techniques and their applications in healthcare. Their review underscores the critical role of watermarking systems in ensuring the authenticity and integrity of medical image data, particularly in the context of telemedicine applications. Furthermore, Tiwari and Srivastava(2024) explored image watermarking techniques based on Schur decomposition and various image invariant moments, shedding light on emerging approaches to watermark embedding and extraction in medical imaging. Their study underscores the importance of leveraging mathematical techniques for enhancing the security and robustness of medical image watermarking systems. Lastly, Al-hussan et al. (2023) proposed hybrid watermarking and encryption techniques for securing medical images, highlighting the interdisciplinary nature of research in medical image security. By integrating watermarking and encryption methods, their approach offers comprehensive protection for medical image data, addressing the evolving threats in digital healthcare environments. Overall, the studies reviewed here collectively contribute to advancing the field of medical image watermarking, offering innovative solutions to address the challenges of data security and integrity in healthcare settings. Balasamy and Shamia (2024) proposed a feature extraction-based medical image watermarking approach using a fuzzy-based median filter, which highlights the significance of incorporating fuzzy logic in image processing for enhancing watermark robustness. Their methodology involves modifying singular values of both clinical and watermark images, demonstrating the effectiveness of feature-based watermark embedding in medical imaging contexts.

Likewise, Fan and colleagues (2023) proposed a multi-watermarking approach tailored for medical images, combining the Inception V3 model with the Discrete Cosine Transform (DCT). Their work highlights the fusion of cutting-edge deep learning architectures with conventional image processing methods.

By leveraging feature extraction methods inherent in Inception V3, their approach offers a comprehensive solution for embedding multiple watermarks in medical images, enhancing data security and integrity. Sivaraman et al. (2023)explored the use of Pullikolam-assisted reconfigurable hardware for medical image watermarking, highlighting the potential of hardware acceleration in enhancing watermarking efficiency and performance. Their study underscores the importance of hardware optimization in facilitating real-time watermark embedding and extraction processes, particularly in resource-constrained environments. Nawaz et al.(2024) proposed a hybrid watermarking algorithm for medical images based on digital transformation and MobileNetV2, introducing a novel approach to watermark embedding and extraction. By combining digital transformation techniques with MobileNetV2 feature extraction, their method offers robustness against various attacks while minimizing computational complexity, making it suitable for real-world applications in healthcare. Benyoucef and Hamaouche (2023) presented a region-based medical image watermarking approach for secure Electronic Patient Record (EPR) transmission in e-Health systems. Their methodology involves embedding watermarks in specific regions of medical images, ensuring data confidentiality and integrity during transmission. This approach addresses the imperceptibility requirement of medical image watermarking, ensuring that watermarks do not affect the clinical interpretation of images. Additionally, Ahmad et al. (2023) utilized the Fuzzy Analytical Hierarchy Process (FAHP) for selecting suitable watermarking techniques for medical image security, highlighting the importance of decision-making methodologies in optimizing watermarking solutions. Their study contributes to the development of a systematic framework for selecting watermarking techniques based on specific security requirements and constraints in medical imaging applications. Devi and Mohapatra (2023) proposed a robust blind medical image watermarking scheme using Grey Wolf Optimized (GWO) Discrete Wavelet Transform (DWT), Discrete Cosine Transform (DCT), and Singular Value Decomposition (SVD), showcasing the potential of metaheuristic optimization in enhancing watermark robustness and imperceptibility. Their methodology involves optimizing watermark embedding parameters using GWO, ensuring robustness against various attacks while preserving image quality. Moreover, Nawaz et al. (2023) introduced a hybrid medical image zero watermarking approach via Discrete Wavelet Transform (DWT)-ResNet101 and Discrete Cosine Transform (DCT), offering a comprehensive solution for zero watermark embedding in medical images. By leveraging deep learning features extracted from ResNet101 and traditional image processing techniques, their method achieves high robustness and imperceptibility, making it suitable for medical image authentication and integrity verification. Finally, Dey et al. (2023) proposed a Singular Value Decomposition (SVD)-based watermarking scheme for medical image authentication, emphasizing the importance of mathematical transformations in ensuring data integrity and authenticity. Their study contributes to the development of secure watermarking techniques tailored specifically for medical imaging applications, addressing the unique security requirements and challenges in healthcare data management. Sinhal et al. (2022)proposed a multipurpose medical image watermarking scheme to provide copyright protection and tamper detection. Their approach involves embedding watermarks into medical images to deter unauthorized use and ensure authenticity. Similarly, Moad et. al. (2022)focused on securing e-healthcare applications through reversible watermarking methods, where watermarked images are encrypted for heightened security. Amine et al. (2023) introduced a blind watermarking scheme tailored for telemedicine applications, emphasizing the importance of preserving image quality while enhancing security. Apostolidis and Papakostas (2022) explored digital watermarking as a potential adversarial attack on medical image analysis with deep learning. Their study sheds light on the dual nature of watermarking, highlighting its potential misuse as a tool for adversarial attacks in the context of deep learning-based medical image analysis. Soualmi et al. (2022) proposed an imperceptible watermarking scheme for medical image tamper detection, emphasizing the need for robust methods to detect unauthorized alterations in medical images. Li et al. (2022) developed a robust multi-watermarking algorithm for medical images using patchwork-DCT, focusing on preserving image quality and robustness against attacks. Singh et al. (2022) introduced Watmif, a multimodal medical image fusion-based watermarking approach for telehealth applications, which involves fusing medical images with cover images and encrypting the marked medical images for secure transmission. Zermi et al.(2022) presented an SVD values ordering scheme for medical image watermarking to ensure the integrity of watermarked images and enhance the robustness of embedded watermarks. Chacko and Chacko (2022) leveraged deep learning and optimization techniques to develop a robust medical image watermarking approach based on DCT and Harris hawks optimization, aiming to enhance watermark robustness and integrity. Ali and Hussain (2022) proposed a novel fragile zero-watermarking algorithm for digital medical images, focusing on preserving image integrity and authenticity. Their approach involves embedding imperceptible watermarks that enable fragile tamper detection without altering image quality. Moad, Kafi, and Khaldi (2022) introduced a wavelet-based medical image watermarking scheme aimed at secure transmission in telemedicine applications. Their methodology involves embedding watermarks into medical images using wavelet transforms and analyzing the robustness of the watermark against common attacks in image watermarking. Xia et al. (2022) developed a robust zero-watermarking algorithm for lossless copyright protection of medical images, addressing the growing concern of digital image piracy. They achieved this by employing a combination of encryption techniques and watermarking based on Zaslavsky chaotic maps and discrete cosine transform (DCT) coefficients. Balaska et al. (2022) proposed a hybrid approach combining encryption and robust blind medical image watermarking, leveraging Zaslavsky chaotic maps and DCT coefficients to enhance security in medical data transmission. These studies highlight the importance of watermarking techniques in safeguarding medical images during transmission and storage in telemedicine applications. In addition to watermarking methods focused on security, there are also efforts to enhance the efficiency and reliability of watermarking schemes for medical images. Magdy et al. (2022) presented multiple zero-watermarking models designed specifically for the Internet of Medical Things (IoMT), aiming to secure medical images transmitted in telemedicine applications using low-cost solutions suitable for embedded systems like Raspberry Pi. Singh et al. (2022) proposed a region-based hybrid medical image watermarking scheme to ensure the authenticity, integrity, and confidentiality of medical images transmitted in IoMT environments. Their approach combines region-based watermarking techniques with hybrid encryption methods to enhance robustness and security in medical data transmission. These studies contribute to the development of watermarking techniques tailored to the unique requirements of telemedicine and IoMT applications, addressing the need for efficient and reliable security measures in the digital transmission of medical images. Furthermore, comprehensive reviews and analyses of digital image watermarking techniques provide valuable insights into the current state of the field and potential avenues for future research. Wadhera et al. (2022) conducted a thorough review of digital image watermarking techniques, focusing on embedding and extraction methods and addressing issues related to security and robustness.

Their review serves as a valuable resource for understanding the various approaches to image watermarking and their applications in different domains, including medical imaging. Alquhayz and Raza (2022)discussed watermarking techniques specifically applied to the security of medical images and image sequences, emphasizing the importance of ensuring the trustworthiness and integrity of medical data in digital environments. These comprehensive reviews provide researchers and practitioners with a deeper understanding of the challenges and opportunities in medical image watermarking, paving the way for further advancements in the field. Reza yatm and and Naghsh(2022) proposed a robust and semi-blind digital image watermarking method based on Discrete Wavelet Transform (DWT) and Singular Value Decomposition (SVD). Their approach aimed to achieve robust watermarking across various types of images, such as pictures and videos, by leveraging DWT and SVD techniques. Alzahrani(2022) introduced an enhanced invisibility and robustness digital image watermarking method based on DWT-SVD. Their method allowed for the embedding and extraction of watermarks without limitations on the host image and watermark information, enhancing both invisibility and robustness. Sanivarapu(2022) developed an adaptive tamper detection watermarking scheme specifically tailored for medical images in the transform domain. This scheme emphasized perceptual transparency and demonstrated superior performance compared to other medical image watermarking techniques. Fan et al.(2022)proposed DwiMark, a multiscale robust deep watermarking framework designed for diffusion-weighted imaging images. Despite limited existing work on deep watermarking for medical images, DwiMark showcased promising results in this domain. Horasan et al. (2022) presented a DWT-SVD-based watermarking technique for high-resolution medical holographic images. Their approach integrated a blind watermarking technique with a digital image signature for authentication, utilizing the discrete wavelet transform. Li et al. (2022) introduced a robust watermarking algorithm for medical images based on the accelerated-KAZE discrete cosine transform (DCT).

Their method extracted feature vectors from medical images and employed zero-watermarking techniques for embedding and extraction. Gupta et al. (2022) proposed a watermarking technique using LSB encryption for DICOM images, focusing on detection and localization of image forgery. Their approach employed visual cryptography-based watermarking, utilizing features, key, and secret shares for forgery detection. Salim et al. (2022) developed a visual cryptography-based watermarking approach for the detection and localization of image forgery, aiming to enhance the security of digital images by detecting and localizing forgeries using secret shares. Gong et. al. (2022) introduced a robust medical zero-watermarking algorithm based on Residual-DenseNet, a deep learning-based approach. Their algorithm demonstrated robustness against geometric attacks and utilized medical images as labels for training. Liu et al. (2022) proposed a robust zero watermarking algorithm for medical images using local binary pattern (LBP) and discrete cosine transform (DCT). Their method combined LBP and DCT to achieve robust watermarking amidst the emergence and use of various digital image processing software. Rezayatmand and Naghsh(2022) proposed a robust and semi-blind digital image watermarking method based on Discrete Wavelet Transform (DWT) and Singular Value Decomposition (SVD). Their approach aimed to achieve robust watermarking across various types of images, such as pictures and videos, by leveraging DWT and SVD techniques. (Fan et al., 2022) introduced an enhanced invisibility and robustness digital image watermarking method based on DWT-SVD. Their method allowed for the embedding and extraction of watermarks without limitations on the host image and watermark information, enhancing both invisibility and robustness. Sanivarapu (2022) developed an adaptive tamper detection watermarking scheme specifically tailored for medical images in the transform domain. This scheme emphasized perceptual transparency and demonstrated superior performance compared to other medical image watermarking techniques.Fan et al.(2022) proposed DwiMark, a multiscale robust deep watermarking framework designed for diffusion-weighted imaging images. Despite limited existing work on deep watermarking for medical images, DwiMark showcased promising results in this domain. Horasan et al. (2022) presented a DWT-SVD-based watermarking technique for high-resolution medical holographic images. Their approach integrated a blind watermarking technique with a digital image signature for authentication, utilizing the discrete wavelet transform. Li et al. (2022) introduced a robust watermarking algorithm for medical images based on the accelerated-KAZE discrete cosine transform (DCT). Their method extracted feature vectors from medical images and employed zero-watermarking techniques for embedding and extraction. Gupta et al. (2022) proposed a watermarking technique using LSB encryption for DICOM images, focusing on detection and localization of image forgery. Their approach employed visual cryptography-based watermarking, utilizing features, key, and secret shares for forgery detection. Salim et al. (2022) developed a visual cryptography-based watermarking approach for the detection and localization of image forgery, aiming to enhance the security of digital images by detecting and localizing forgeries using secret shares. Gong et al. (2022) introduced a robust medical zero-watermarking algorithm based on Residual-DenseNet, a deep learning-based approach. Their algorithm demonstrated robustness against geometric attacks and utilized medical images as labels for training. Begum and Uddin (2020) conducted a review of digital image watermarking techniques, focusing on their application in e-governance. The methodology involved an extensive literature search to identify relevant studies and analyze their methodologies, achievements, and applications in the context of e-governance. Kumar et al. (2020) provided a review of digital image watermarking techniques using LSB and DWT. The methodology involved examining the background of these techniques, including their principles and applications. The review analyzed existing literature to highlight the implementation of LSB and DWT-based watermarking methods in sustainable development contexts. Lee, et al. (2020) proposed a CNN-based digital image watermarking method adaptive to image and watermark resolutions. The methodology involved developing a convolutional neural network model capable of dynamically adjusting to varying resolutions of host images and watermarks. The proposed approach was evaluated through experiments to assess its effectiveness. Sinhal and Ansari (2020) introduced a multipurpose image watermarking scheme for digital image protection. The methodology included the development of a watermarking technique combining robust and fragile watermarking methods. The proposed approach was evaluated through simulations to demonstrate its applicability and effectiveness for authentication and security purposes. Evsutin et al. (2020) reviewed current research directions in digital steganography and watermarking for digital images. The methodology involved synthesizing recent advancements in these fields by analyzing a wide range of literature sources. The review identified emerging trends and future research directions based on the analysis of existing studies. Sinha et al. (2020) proposed a spatial domain-based digital image watermarking scheme for copyright protection. The methodology included the development of a watermarking algorithm designed to embed copyright information into digital images in the spatial domain. The proposed approach was evaluated through simulations to assess its effectiveness in protecting copyright. Mohanarathinam et al.(2020) undertook an extensive examination of digital watermarking techniques aimed at enhancing image security. Their methodology encompassed a thorough analysis of various watermarking methods, evaluating their advantages and drawbacks through an extensive review of existing literature. The objective was to offer a comprehensive understanding of the strengths and weaknesses associated with different watermarking approaches. Hatoum et al. (2021) evaluated the use of deep learning for image watermarking attack. The methodology involved assessing the effectiveness of deep learning models in attacking various watermarking techniques. The study conducted experiments to evaluate the vulnerability of existing watermarking methods to deep learning-based attacks. Deeba et al.(2020) proposed a digital image watermarking method based on ANN and LSB. The methodology included the development of an algorithm using Artificial Neural Networks (ANN) for watermark embedding and extraction in the LSB domain. The proposed approach was evaluated through simulations to assess its performance and limitations.

## 3. THE PROPOSED METHOD

The employed methodology aimed to enhance the robustness and image quality of watermarking through a combination of discrete wavelet transform (DWT), singular value decomposition (SVD), and set partitioning in hierarchical trees (SPIHT). Initially, the cover image underwent second level DWT, followed by SVD on the obtained DWT coefficients (Kumar, 2024). To reinforce the security of the logo watermark, Arnold encryption was applied, resulting in two sub-images of size 128×128 (Kumar, 2023). These sub-images were then embedded into the 'LH' and 'HL' sub-bands of the 1st level DWT of the cover image, while the text watermark, encoded using hamming code, was embedded into the 'HH' sub-bands of the second level DWT. The watermarked image underwent inverse SVD and DWT, followed by SPIHT compression.

In the extraction process, the aforementioned steps were reversed to retrieve the original logo and text watermarks. Experimental evaluations compared the performance of the proposed method, integrating DWT, DCT, and SPIHT techniques, against various watermarking attacks, including noise addition, compression, filtering, and cropping. The results demonstrated the method's superiority in robustness and image quality, exhibiting better trade-offs compared to existing methods. Notably, the proposed method maintained high robustness against attacks while preserving cover image quality.

Leveraging Particle Swarm Optimization and the Butterfly Optimization Algorithm techniques, the methodology utilized these optimization algorithms to enhance the watermarking process's effectiveness and efficiency. The combination of DWT, SVD, and SPIHT, augmented by optimization techniques, contributed to the proposed method's resilience against attacks and its ability to maintain image quality. The experiments validated the effectiveness and practical utility of the proposed method for diverse applications, including copyright protection, tamper detection, and authentication, underscoring its promising real-world applicability.

## 4. EMBEDDING ALGORITHM

The embedding algorithm utilized a novel approach to enhance the robustness and integrity of watermarking in digital images. Here's a summary of the steps involved:

1. Hybrid Optimization Technique: The cover image and Watermark 1 underwent optimization using a hybrid optimization technique combining whale and cuckoo optimization methods. This process aimed to improve the quality of the cover image and ensure efficient embedding of Watermark 1 while maintaining its integrity.
2. Discrete Wavelet Transform (DWT): The optimized cover image was decomposed using DWT to obtain four sub-bands (LL, LH, HL, HH). DWT facilitates a multi-resolution representation, allowing selective embedding of watermarks in specific frequency components and spatial details.
3. Selection of Sub-bands: Two distinct sub-bands (LH and HH) were selected for embedding Watermark 1 and Watermark 2, respectively, to enhance resistance against different image processing attacks.
4. Encryption and Encoding: Watermark 1 underwent encryption using the Arnold transform, while Watermark 2 was encoded using a Hamming encoder. These processes added a layer of security to the watermarks, preventing unauthorized access and tampering.
5. Optimal Scaling Factors: The algorithm calculated optimal scaling factors for each sub-band, determining the strength of the watermark in each sub-band during embedding to achieve a balance between robustness and visual quality.
6. Embedding Watermarks: Watermarks were embedded in their respective sub-bands using the obtained scaling factors, followed by the application of Discrete Cosine Transform (DCT) to increase imperceptibility.
7. Compression: Watermarked sub-bands were compressed using the Set Partitioning in Hierarchical Trees (SPIHT) algorithm to efficiently store and transmit the watermarked image while preserving watermark information.
8. Encryption: Compressed watermarked image is encrypted using chaotic encryption technique.
9. Reconstruction: The watermarked image was reconstructed using the inverse DWT, allowing for visualization and analysis. Quality metrics were applied to evaluate robustness against various attacks, such as JPEG compression and noise addition.
10. Evaluation: A quality metric was applied to assess the algorithm's robustness against watermarking attacks, and the results, along with the watermarked image, were outputted for analysis.

The extraction algorithm outlined a series of steps to accurately retrieve Watermark 1 and Watermark 2 from the watermarked image.

## 5. EXTRACTION ALGORITHM

Certainly, here's the rewritten extraction algorithm without changing the equations:

1. Decomposition using Discrete Wavelet Transform (DWT):
   - Decompose the watermarked image using DWT into four sub-bands: LL, LH", HL", and HH". This step separates the image into different frequency components.
2. Selection of Embedded Sub-bands:
   - Select the sub-bands LH" and HH" containing Watermark 1 and Watermark 2, respectively, as they were embedded during the encoding process.
3. Inverse Discrete Cosine Transform (IDCT):
   - Apply IDCT to LH" and HH" sub-bands to retrieve the spatial domain representations of the embedded watermarks, ew1' and ew2'.
4. Obtaining Scaling Factors:
   - Obtain the scaling factors (sf_LH and sf_HH) used during embedding to adjust the strength of Watermark 1 and Watermark 2.
5. Decryption and Decoding:
   - Decrypt Watermark 1 using the inverse Arnold transform to obtain the original watermark, w1. Decode Watermark 2 using a Hamming decoder to correct any errors introduced during encoding.
6. Quality Metric Evaluation:
   - Assess the accuracy of the extracted watermarks compared to the originals using a quality metric, ensuring fidelity and reliability.
7. Decryption:
   - Compressed watermarked image is decrypted using chaotic decryption technique.
8. Decompression using SPIHT Algorithm:
   - Decompress the compressed LH" and HH" sub-bands using the SPIHT algorithm to recover the original sub-bands.

   8. Inverse DCT:

   - Reconstruct the original watermarked sub-bands, LH'(x,y) and HH'(x,y), by applying IDCT to the decompressed LH" and HH" sub-bands.
9. Watermark Extraction:

- Utilize the obtained scaling factors to extract the watermarks, ew1(x,y) and ew2(x,y), from the watermarked sub-bands, ensuring accuracy and integrity.

By following these steps, the extraction algorithm ensures the accurate retrieval of Watermark 1 and Watermark 2 from the watermarked image, facilitating various applications such as copyright protection and authentication.

This comprehensive approach to watermark embedding ensures both security and integrity, making it suitable for various applications requiring copyright protection and authentication of digital images.

## 6. EXPERIMENTAL RESULTS

The proposed watermarking method underwent rigorous evaluation using MATLAB version 2013b, employing diverse cover images with dimensions of 512×512 pixels (Alhussan et al., 2023) and two types of watermarks: a 'logo' sized 256×256 pixels and a 'secret code' watermark consisting of 96 bits (Shivani & Shenapati, 2017). Evaluation involved subjecting the proposed scheme to various attacks to assess its performance comprehensively. Standard metrics including Peak Signal to Noise Ratio (PSNR), Normalized Correlation (NC), Bit Error Rate (BER), and Structured Similarity Index (SSIM) were utilized for performance evaluation, as outlined in Figure 1. Figure 2 illustrates the considered cover image, the two watermarks, and the resulting watermarked images. Subsequently, Figure 3 displays the extracted logo and 'secret code' watermarks, while Figure 4 depicts the extracted watermarks alongside watermarked (attacked) images. Detailed performance analysis is presented in Tables 1 through 9. Figure demonstrated the comparison of proposed scheme against other mentioned techniques.

The experimental results underscore the remarkable robustness of the proposed method against diverse attacks, evident from the high PSNR, NC, and SSIM values, coupled with low BER values. Notably, Table 1 showcases the method's performance across varying gain and bit rates, with the highest PSNR recorded at 42 dB for a gain factor of 0.01 and bit rate of 3.5. Conversely, the lowest PSNR of 31.35 dB was observed at the same gain and bit rate of 0.1. The evaluation also revealed exceptional NC and SSIM values across different settings, underscoring the method's resilience. Table 2 further explores the performance using whale optimization and cuckoo search algorithms, showcasing significant improvements in PSNR,SSIM and NC values. Moreover, Table 3 examines the scheme's performance under fixed gain values, demonstrating consistent high-quality results across different metrics. The evaluation encompassed a variety of cover images, as summarized in Table 4, elucidating the method's adaptability and efficacy across diverse scenarios. Additionally, Table 5 delves into the method's robustness against geometrical and signal processing attacks, highlighting its superiority in maintaining high NC values even under adverse conditions. Table 6 demonstrate test results against different attacks. Best value for NC is 0.9999.Furthermore; Table 7 elucidates the impact of character string size on performance, with the method consistently achieving impressive results across various configurations. The comparison outlined in Tables 8,Table 9 showcase the performance of proposed scheme against other in term of NC, PSNR and SSIM values. However Table 10, Table 11 highlights the dominant performance of the proposed approach compared to established methods in term of NPCR and UACI values, especially regarding the Normalized Correlation (NC) values across various attack scenarios.

The method's effectiveness is further emphasized in Figure 5, depicting the robust performance of the proposed scheme under various optimization algorithms, with the hybrid algorithm yielding the most promising results across all evaluated metrics. Overall, the experimental findings affirm the efficacy and resilience of the proposed scheme, positioning it as a viable solution for safeguarding digital content against diverse threats.

*Table 1. Experimental results for bit rate=1.5,2.5,3.5 and different gain factors*

| Gain factors (μ) | Bit rate Values(SPIHT) | Gain factors (μ) | PSNR (dB) | NC | BER | SSIM values |
|---|---|---|---|---|---|---|
| 0.01 | 1.5 | 0.01 | 41.63 | 0.8789 | 0 | 0.9998 |
| 0.03 | | 0.04 | 39.15 | 0.9845 | 0 | 0.9969 |
| 0.06 | | 0.06 | 35 | 0.9836 | 0 | 0.9976 |
| 0.1 | | 0.1 | 31.35 | 0.9978 | 0 | 0.9878 |
| 0.01 | 2.5 | 0.01 | 41.95 | 0.8918 | 0 | 0.9957 |
| 0.04 | | 0.04 | 39.58 | 0.9945 | 0 | 0.9968 |
| 0.06 | | 0.06 | 35.66 | 0.9955 | 0 | 0.9958 |
| 0.1 | | 0.1 | 31.66 | 0.9975 | 0 | 0.9875 |
| 0.01 | 3.5 | 0.01 | 42 | 0.8936 | 0 | 0.9989 |
| 0.04 | | 0.04 | 39.59 | 0.9940 | 0 | 0.9979 |
| 0.06 | | 0.06 | 35.63 | 0.9951 | 0 | 0.9949 |
| 0.1 | | 0.1 | 31.67 | 0.9983 | 0 | 0.9886 |

*Table 2. Experimental results using various optimization techniques and different gain factors*

| Techniques | Gain factors (μ) | PSNR (dB) | NC | BER | SSIM values |
|---|---|---|---|---|---|
| using Whale optimization | 0.01 | 43.63 | 0.9804 | 0 | 0.9978 |
| | 0.04 | 40.25 | 0.9825 | 0 | 0.9967 |
| | 0.06 | 37 | 0.9829 | 0 | 0.9956 |
| | 0.1 | 35.35 | 0.9835 | 0 | 0.9874 |
| cuckoo search algorithm | 0.01 | 42.95 | 0.8937 | 0 | 0.9954 |
| | 0.04 | 40.58 | 0.9814 | 0 | 0.9969 |
| | 0.06 | 37.66 | 0.9978 | 0 | 0.9959 |
| | 0.1 | 33.66 | 0.9985 | 0 | 0.9875 |

*Table 3. Test results at fixed gain(0.01)*

| Bit rate | PSNR | NC | BER | SSIM |
|---|---|---|---|---|
| 1 | 43.73 | 0.9970 | 0 | 0.9949 |
| 2 | 42.97 | 0.9785 | 0 | 0.9944 |
| 3 | 42 | 0.97887 | 0 | 0.9945 |

*Table 4. Test results for different covers images (against gain 0.04 & bit rate 3)*

| Cover images | PSNR (in (dB) | NC | BER | NPCR | UACI | SSIM | CR |
|---|---|---|---|---|---|---|---|
| MRI | 43.09 | 0.9998 | 0 | 0.9980 | 0.3999 | 0.9973 | 0.1245 |
| Moon | 42.48 | 0.9998 | 0 | 0.9978 | 0.4395 | 0.9996 | 0.0940 |
| Tire | 45.81 | 0.9998 | 0 | 0.9979 | 0.3951 | 0.9988 | 0.0252 |
| Cameraman | 37.24 | 0.9993 | 0 | 0.9979 | 0.3651 | 0.9985 | 0.0347 |
| Coins | 42.32 | 0.9998 | 0 | 0.9965 | 0.3186 | 0.9988 | 0.0348 |
| Boat | 35.27 | 0.9991 | 0 | 0.9973 | 0.2949 | 0.9939 | 0.1257 |
| Finger | 34.39 | 0.9940 | 0 | 0.9961 | 0.2889 | 0.9829 | 0.0327 |
| Bird | 51.33 | 0.9997 | 0 | 0.9946 | 0.2892 | 0.9989 | 0.0531 |
| Barbara | 33.01 | 0.9989 | 0 | 0.9961 | 0.2980 | 0.9924 | 0.1246 |
| Baboon | 39.24 | 0.9998 | 0 | 0.9982 | 0.3296 | 0.9896 | 0.0325 |

*Table 5. Performance of the proposed method for changing gain values*

| Gain factors | PSNR (in dB) | NC | BER | NPCR | UACI | SSIM | CR |
|---|---|---|---|---|---|---|---|
| 0.01 | 38.09 | 0.9862 | 0 | 0.9962 | 0.2861 | 0.9975 | 0.1364 |
| 0.04 | 38.08 | 0.9988 | 0 | 0.9961 | 0.2859 | 0.9974 | 0.1367 |
| 0.08 | 37.07 | 0.999 | 0 | 0.9961 | 0.2867 | 0.9973 | 0.1364 |
| 0.1 | 37.05 | 0.9997 | 0 | 0.9960 | 0.2864 | 0.9973 | 0.1365 |
| 0.4 | 38.02 | 0.9999 | 0 | 0.9960 | 0.2863 | 0.9972 | 0.1362 |

*Table 6. Test results against different attacks.*

| Attacks applied | NC values | BER | SSIM |
|---|---|---|---|
| Salt & pepper noise attack(with density=0.01,0.08,0.001) | 0.7877,0.6875, 0.9984 | 0 | 0.9582,0.9562,0.9953 |
| Gaussian noise-attack(mean=0,variance=0.01,0.001,0.002) | 0.5965,0.7381,0.6275 | 0 | 0.8943,0.9832,0.9778 |
| Median filter[2 2],[3 3] | 0.8891,0.8762 | 0.9532 | 0.9958,0.9976 |
| JPEG compression (QF=50& QF=80) | 0.9991& 0.9989 | 0 | 0.9947& 0.9946 |
| Histogram equalization | 0.8247 | 1.3276 | 0.4583 |
| Cropping(20 20 400 480) | 0.7691 | 2.36 | 0.3090 |
| Speckle noise attack(at density=0.01, 0.02 &0.005) | 0.8721, 0.8919 & 0.8919 | 0 | 0.9867, 0.98840 &0.9899 |
| Average filtering | 0.9999 | 0.9885 | 0.9952 |

*Table 7. Performance under (24 and15) characters under bit rate 3 and different gain.*

| Gain factor(μ) | 24 characters | | | | 15 characters | | | |
|---|---|---|---|---|---|---|---|---|
| | PSNR | NC | BER | SSIM | PSNR | NC | BER | SSIM |
| 0.01 | 42.7071 | 0.9710 | 0 | 0.9988 | 40.72 | 0.9570 | 0 | 0.9987 |
| 0.04 | 39.1801 | 0.9980 | 0 | 0.9979 | 38.61 | 0.9990 | 0 | 0.9955 |
| 0.1 | 37.3564 | 0.9999 | 0 | 0.9896 | 35.49 | 0.9975 | 0 | 0.9968 |

*Table 8. Comparison results of our scheme with other author*

| Attacks | Alhussan et al.(2023) | Proposed scheme |
|---|---|---|
| Salt and pepper noise(0.02) | 0.9999 | 0.9999 |
| Gaussian noise(0,0.01) | 0.9999 | 0.9999 |
| Speckle noise (0.02) | - | 0.9911 |
| Gaussian low pass filter (3*3) | - | 0.9999 |
| Average filter | - | 0.9984 |
| Median filter[3 3] | 0.97 | 0.9997 |
| Rotation | 0.85 | 0.9998 |
| Histogram equalization | - | 0.9997 |
| Sharpening | - | 0.9756 |
| Wiener filter (11*11) | - | 0.9998 |
| JPEG (90) | - | 0.9999 |
| JPEG 2000 (10) | - | 1 |
| Dither | - | 0.7889 |
| Motion blur | - | 1 |
| Shearing | - | 0.9999 |
| Cropping | 0.98 | 0.9999 |

*Table 9. Comparing PSNR, SSIM and NC of proposed method with other compared methods.*

| Cover image | Bit Rate (bpp) | Methods | PSNR(in dB) | SSIM | NC |
|---|---|---|---|---|---|
| Lena | 0.25 | Shivani and senapati(2017) | 30.93 | 0.8982 | 0.3201 |
| | | Said et al.(1996) | 31.24 | 0.9075 | 0.1317 |
| | | Our | 32.73 | 0.9929 | 0.9705 |
| | 0.5 | Shivani and senapati(2017) | 31.36 | 0.9292 | 0.9804 |
| | | Said et al.(1996) | 31.10 | 0.9167 | 0.9848 |
| | | Our | 35.03 | 0.9959 | 0.9911 |
| | 1 | Shivani and senapati(2017) | 31.13 | 0.9617 | 0.9804 |
| | | Said et al.(1996) | 33.87 | 0.9600 | 0.9801 |
| | | Our | 36.20 | 0.9970 | 0.9892 |
| | 1.5 | Shivani and senapati(2017) | 34.40 | 0.9676 | 0.9149 |
| | | Said et al.(1996) | 34.35 | 0.9658 | 0.9227 |
| | | Our | 36.54 | 0.9972 | 0.9885 |

continued on following page

*Table 9. Continued*

| | | | | | |
|---|---|---|---|---|---|
| | 2 | Shivani and senapati(2017) | 34.90 | 0.9693 | 0.9257 |
| | | Said et al.(1996) | 34.83 | 0.9686 | 0.9298 |
| | | Our | 36.58 | 0.9973 | 0.9896 |
| | 2.5 | Shivani and senapati(2017) | 35.44 | 0.9740 | 0.9598 |
| | | Said et al.(1996) | 35.41 | 0.9738 | 0.9510 |
| | | Our | 36.70 | 0.9973 | 0.9896 |
| | 3 | Shivani and senapati(2017) | 35.65 | 0.9739 | 0.9558 |
| | | Said et al.(1996) | 35.65 | 0.9724 | 0.9490 |
| | | Our | 36.72 | 0.9974 | 0.9902 |
| | Lossless at 3.5 | Shivani and senapati(2017) | 36.00 | 0.9750 | 0.9831 |
| | | Said et al.(1996) | 35.83 | 0.9762 | 0.9841 |
| | | Our | 36.77 | 0.9975 | 0.9903 |

*Table 10. NPCR value comparison between proposed and compared technique (Scaling factor optimization)*

| **Image** | **Wang et al.(2015)** | **Belazi et al.(2016)** | **Wang et al.(2012)** | **Hua et al.(2015)** | **Proposed method** |
|---|---|---|---|---|---|
| Barbara | 99.4285 | 99.5227 | 99.6092 | 99.6162 | 99.665 |
| Boat | 99.4509 | 99.5609 | 99.6102 | 99.6281 | 99.672 |
| Cameraman | 99.7335 | 99.5749 | 99.6205 | 99.6292 | 99.7337 |
| Lena | 99.5177 | 99.5511 | 99.6228 | 99.6146 | 99.677 |
| Peppers | 99.5154 | 99.5808 | 99.6319 | 99.6092 | 99.734 |

*Table 11. UACI value comparison between proposed and compared technique (Scaling factor optimization)*

| **Image** | **Wang et al.(2015)** | **Belazi et al.(2016)** | **Wang et al.(2012)** | **Hua et al.(2015)** | **Proposed method** |
|---|---|---|---|---|---|
| Barbara | 33.6177 | 33.3890 | 33.7431 | 33.5776 | 33.744 |
| Boat | 32.4296 | 33.4176 | 33.5367 | 33.6143 | 33.705 |
| Cameraman | 32.5297 | 33.3691 | 33.7786 | 33.7050 | 34.751 |
| Lena | 33.2231 | 33.3461 | 33.7041 | 33.5561 | 33.981 |
| Peppers | 33.2263 | 33.3540 | 33.6923 | 33.6284 | 33.972 |

## CONCLUSION

Our proposed chapter not only introduces an advanced watermarking method for medical images but also delves into the technical intricacies, security considerations, and real-world applicability of the proposed solution. By addressing the unique challenges posed by medical imaging data, our method offers a secure, robust, and practical solution for safeguarding critical healthcare assets. The integration of hybrid optimization techniques, encryption strategies, and rigorous testing against real-world challenges positions our proposed method as a valuable contribution to the evolving landscape of medical data security. Through this comprehensive analysis, we envision our proposed method as an asset with far-reaching applications, including copyright protection, authentication, and secure data transmission within healthcare ecosystems.

## REFERENCES


Ahmad, M., Nadeem, M., Islam, M., Ali, S., Agrawal, A., & Ahmad Khan, R. (2023). Selection of Digital Watermarking Techniques for Medical Image Security by Using the Fuzzy Analytical Hierarchy Process. Recent Advances in Computer Science and Communications (Formerly: Recent Patents on Computer Science), 16(7), 74-80.

Alhussan, A. A., Abdallah, H. A., Alsodairi, S., & Ateya, A. A. (2023). Hybrid Watermarking and Encryption Techniques for Securing Medical Images. *Computer Systems Science and Engineering*, *46*(1), 403–416. DOI: 10.32604/csse.2023.035048



Alhussan, A. A., Abdallah, H. A., Alsodairi, S., & Ateya, A. A. (2023). Hybrid Watermarking and Encryption Techniques for Securing Medical Images. *Computer Systems Science and Engineering*, *46*(1), 403–416. DOI: 10.32604/csse.2023.035048



Ali, Z., & Hussain, M. (2022). A novel fragile zero-watermarking algorithm for digital medical images. *Electronics (Basel)*, *11*(5), 710. DOI: 10.3390/electronics11050710

Alquhayz, H., & Raza, B. (2022). Watermarking techniques for the security of medical images and image sequences. *Arabian Journal for Science and Engineering*, *47*(8), 9471–9488. DOI: 10.1007/s13369-021-06254-7

Alzahrani, A. (2022). Enhanced invisibility and robustness of digital image watermarking based on DWT-SVD. *Applied Bionics and Biomechanics*, *2022*, 2022. DOI: 10.1155/2022/5271600 PMID: 35126659



Alzahrani, A. (2022). Enhanced invisibility and robustness of digital image watermarking based on DWT-SVD. *Applied Bionics and Biomechanics*, *2022*, 2022. DOI: 10.1155/2022/5271600 PMID: 35126659



Amine, K., Fares, K., Redouane, K. M., & Salah, E. (2022). Medical image watermarking for telemedicine application security. *Journal of Circuits, Systems, and Computers*, *31*(05), 2250097. DOI: 10.1142/S0218126622500979

Apostolidis, K. D., & Papakostas, G. A. (2022). Digital watermarking as an adversarial attack on medical image analysis with deep learning. *Journal of Imaging*, *8*(6), 155. DOI: 10.3390/jimaging8060155 PMID: 35735954

Balasamy, K., & Shamia, D. (2023). Feature extraction-based medical image watermarking using fuzzy-based median filter. *Journal of the Institution of Electronics and Telecommunication Engineers*, *69*(1), 83–91. DOI: 10.1080/03772063.2021.1893231

Balaska, N., Belmeguenai, A., Goutas, A., Ahmida, Z., & Boumerdassi, S. (2022). Securing Medical Data by Combining Encryption and Robust Blind Medical Image Watermarking Based on Zaslavsky Chaotic Map and DCT Coefficients. *SN Computer Science*, *3*(2), 118. DOI: 10.1007/s42979-021-01012-w

Begum, M., & Uddin, M. S. (2020). Digital image watermarking techniques: A review. *Information (Basel)*, *11*(2), 110. DOI: 10.3390/info11020110

Begum, M., & Uddin, M. S. (2020). Analysis of digital image watermarking techniques through hybrid methods. *Advances in Multimedia*, *2020*, 1–12. DOI: 10.1155/2020/7912690

Belazi, A., Abd El-Latif, A. A., & Belghith, S. (2016). A novel image encryption scheme based on substitution-permutation network and chaos. *Signal Processing*, *128*, 155–170. DOI: 10.1016/j.sigpro.2016.03.021

Benyoucef, A., & Hamaouche, M. H. (2023). Region-based medical image watermarking approach for secure EPR transmission applied to e-Health. *Arabian Journal for Science and Engineering*, **•••**, 1–13.

Chacko, A., & Chacko, S. (2022). Deep learning based robust medical image watermarking exploiting DCT and Harris hawks optimization. *International Journal of Intelligent Systems*, *37*(8), 4810–4844. DOI: 10.1002/int.22742

. Deeba, F., Kun, S., Dharejo, F. A., & Memon, H. (2020). Digital image watermarking based on ANN and least significant bit. Information Security Journal: A Global Perspective, 29(1), 30-39.

Devi, H. S., & Mohapatra, H. (2023). A novel robust blind medical image watermarking using GWO optimized DWT-DCT-SVD. *Multimedia Tools and Applications*, *82*(26), 41267–41286. DOI: 10.1007/s11042-023-15158-7

Dey, A., Chowdhuri, P., Pal, P., & Tzu-Chuen, L. (2023, March). SVD-Based Watermarking Scheme for Medical Image Authentication. In International Conference on Network Security and Blockchain Technology (pp. 135-146). Singapore: Springer Nature Singapore.

Evsutin, O., Melman, A., & Meshcheryakov, R. (2020). Digital steganography and watermarking for digital images: A review of current research directions. *IEEE Access : Practical Innovations, Open Solutions*, *8*, 166589–166611. DOI: 10.1109/ACCESS.2020.3022779

Fan, B., Li, Z., & Gao, J. (2022). DwiMark: A multiscale robust deep watermarking framework for diffusion-weighted imaging images. *Multimedia Systems*, *28*(1), 295–310. DOI: 10.1007/s00530-021-00835-0

Fan, B., Li, Z., & Gao, J. (2022). DwiMark: A multiscale robust deep watermarking framework for diffusion-weighted imaging images. *Multimedia Systems*, *28*(1), 295–310. DOI: 10.1007/s00530-021-00835-0

Fan, Y., Li, J., Bhatti, U. A., Shao, C., Gong, C., Cheng, J., & Chen, Y. (2023). A multi-watermarking algorithm for medical images using inception v3 and dct. *CMC-Computers Materials & Continua*, *74*(1), 1279–1302. DOI: 10.32604/cmc.2023.031445

Gong, C., Liu, J., Gong, M., Li, J., Bhatti, U. A., & Ma, J. (2022). Robust medical zero-watermarking algorithm based on Residual-DenseNet. *IET Biometrics*, *11*(6), 547–556. DOI: 10.1049/bme2.12100

Gong, C., Liu, J., Gong, M., Li, J., Bhatti, U. A., & Ma, J. (2022). Robust medical zero watermarking algorithm based on Residual-DenseNet. *IET Biometrics*, *11*(6), 547–556. DOI: 10.1049/bme2.12100

Gull, S., & Parah, S. A. (2024). Advances in medical image watermarking: A state of the art review. *Multimedia Tools and Applications*, *83*(1), 1407–1447. DOI: 10.1007/s11042-023-15396-9 PMID: 37362709

Gupta, M. K., Dadheech, P., Kumar, A., Dogiwal, S. R., Poonia, R. C., Raja, L., & Bhatt, D. P. (2022). Detection and localization for watermarking technique using LSB encryption for DICOM Image. *Journal of Discrete Mathematical Sciences and Cryptography*, *25*(1), 193–204. DOI: 10.1080/09720529.2021.2009193

Gupta, M. K., Dadheech, P., Kumar, A., Dogiwal, S. R., Poonia, R. C., Raja, L., & Bhatt, D. P. (2022). Detection and localization for watermarking technique using LSB encryption for DICOM Image. *Journal of Discrete Mathematical Sciences and Cryptography*, *25*(1), 193–204. DOI: 10.1080/09720529.2021.2009193

Hatoum, M. W., Couchot, J. F., Couturier, R., & Darazi, R. (2021). Using deep learning for image watermarking attack. *Signal Processing Image Communication*, *90*, 116019. DOI: 10.1016/j.image.2020.116019

Horasan, F., Ali Pala, M., Durdu, A., Akgül, A., Akmeşe, Ö. F., & Yıldız, M. Z. (2022). DWT-SVD based watermarking for high-resolution medical holographic images. *Complexity*, *2022*(1), 2022. DOI: 10.1155/2022/3154650

Horasan, F., Ali Pala, M., Durdu, A., Akgül, A., Akmeşe, Ö. F., & Yıldız, M. Z. (2022). DWT-SVD based watermarking for high-resolution medical holographic images. *Complexity*, *2022*(1), 2022. DOI: 10.1155/2022/3154650

Hua, Z., Zhou, Y., Pun, C. M., & Chen, C. P. (2015). 2D Sine Logistic modulation map for image encryption. *Information Sciences, 297*, 80–94. DOI: 10.1016/j.ins.2014.11.018

Kumar, A. (2020). A review on implementation of digital image watermarking techniques using LSB and DWT. Information and Communication Technology for Sustainable Development: Proceedings of ICT4SD 2018, 595-602.

Kumar, C. (2023). Hybrid optimization for secure and robust digital image watermarking with DWT, DCT and SPIHT. *Multimedia Tools and Applications, 83*(11), 1–22. DOI: 10.1007/s11042-023-16903-8

Kumar, C. (2024). Optimized and secure digital image watermarking technique using Henon mapping in redundant domain. *Multimedia Tools and Applications, 83*(34), 1–16. DOI: 10.1007/s11042-024-18643-9

Kumar, C., Singh, A. K., & Kumar, P. (2018). A recent survey on image watermarking techniques and its application in e-governance. *Multimedia Tools and Applications, 77*(3), 3597–3622. DOI: 10.1007/s11042-017-5222-8

Lee, J. E., Seo, Y. H., & Kim, D. W. (2020). Convolutional neural network-based digital image watermarking adaptive to the resolution of image and watermark. *Applied Sciences (Basel, Switzerland), 10*(19), 6854. DOI: 10.3390/app10196854

Li, D., Chen, Y. W., Li, J., Cao, L., Bhatti, U. A., & Zhang, P. (2022). Robust watermarking algorithm for medical images based on accelerated-KAZE discrete cosine transform. *IET Biometrics, 11*(6), 534–546. DOI: 10.1049/bme2.12102

Li, Y., Li, J., Shao, C., Bhatti, U. A., & Ma, J. (2022, July). Robust multi-watermarking algorithm for medical images using patchwork-DCT. In *International Conference on Artificial Intelligence and Security* (pp. 386-399). Cham: Springer International Publishing. DOI: 10.1007/978-3-031-06791-4_31

Liu, W., Li, J., Shao, C., Ma, J., Huang, M., & Bhatti, U. A. (2022, July). Robust zero watermarking algorithm for medical images using local binary pattern and discrete cosine transform. In *International Conference on Artificial Intelligence and Security* (pp. 350-362). Cham: Springer International Publishing. DOI: 10.1007/978-3-031-06764-8_28

Magdy, M., Ghali, N. I., Ghoniemy, S., & Hosny, K. M. (2022). Multiple zero-watermarking of medical images for internet of medical things. *IEEE Access : Practical Innovations, Open Solutions, 10*, 38821–38831. DOI: 10.1109/ACCESS.2022.3165813

Moad, M. S., Kafi, M. R., & Khaldi, A. (2022). Medical image watermarking for secure e-healthcare applications. *Multimedia Tools and Applications*, *81*(30), 44087–44107. DOI: 10.1007/s11042-022-12004-0

Moad, M. S., Kafi, M. R., & Khaldi, A. (2022). A wavelet based medical image watermarking scheme for secure transmission in telemedicine applications. *Microprocessors and Microsystems*, *90*, 104490. DOI: 10.1016/j.micpro.2022.104490

Mohanarathinam, A., Kamalraj, S., Prasanna Venkatesan, G. K. D., Ravi, R. V., & Manikandababu, C. S. (2020). Digital watermarking techniques for image security: A review. *Journal of Ambient Intelligence and Humanized Computing*, *11*(8), 3221–3229. DOI: 10.1007/s12652-019-01500-1

Nawaz, S. A., Li, J., Bhatti, U. A., Shoukat, M. U., Li, D., & Raza, M. A. (2024). Hybrid watermarking algorithm for medical images based on digital transformation and MobileNetV2. *Information Sciences*, *653*, 119810. DOI: 10.1016/j.ins.2023.119810

Nawaz, S. A., Li, J., Shoukat, M. U., Bhatti, U. A., & Raza, M. A. (2023). Hybrid medical image zero watermarking via discrete wavelet transform-ResNet101 and discrete cosine transform. *Computers & Electrical Engineering*, *112*, 108985. DOI: 10.1016/j.compeleceng.2023.108985

Pallaw, V. K., Singh, K. U., Kumar, A., Singh, T., Swarup, C., & Goswami, A. (2023). A robust medical image watermarking scheme based on nature-inspired optimization for telemedicine applications. *Electronics (Basel)*, *12*(2), 334. DOI: 10.3390/electronics12020334

Rezayatmand, M., & Naghsh, A. (2022). Robust and semi-blind digital image watermarking method based on DWT and SVD. *Journal of Intelligent Procedures in Electrical Technology*, *13*(51), 1–18.

. Saber, M., El-Kenawy, E. S. M., Ibrahim, A., & Eid, M. M. (2023). Watermarking System for Medical Images Using Optimization Algorithm. Fusion: Practice & Applications, 10(1).

Said, A., & Pearlman, W. A. (1996). A new, fast, and efficient image codec based on set partitioning in hierarchical trees. *IEEE Transactions on Circuits and Systems for Video Technology*, *6*(3), 243–250. DOI: 10.1109/76.499834

Salim, M. Z., Abboud, A. J., & Yildirim, R. (2022). A visual cryptography-based watermarking approach for the detection and localization of image forgery. *Electronics (Basel)*, *11*(1), 136. DOI: 10.3390/electronics11010136

Sanivarapu, P. V. (2022). Adaptive tamper detection watermarking scheme for medical images in transform domain. *Multimedia Tools and Applications*, *81*(8), 11605–11619. DOI: 10.1007/s11042-022-12273-9

Sheng, M., Li, J., Bhatti, U. A., Liu, J., Huang, M., & Chen, Y. W. (2023). Zero watermarking algorithm for medical image based on Resnet50-DCT. *Computers, Materials & Continua*, *75*(1), 293–309. DOI: 10.32604/cmc.2023.036438

Shivani, J. D., & Senapati, R. K. (2017). Robust image embedded watermarking using DCT and listless SPIHT. *Future Internet*, *9*(3), 33. DOI: 10.3390/fi9030033

Singh, K. N., Singh, O. P., Singh, A. K., & Agrawal, A. K. (2022). Singh, K. N., Singh, O. P., Singh, A. K., & Agrawal, A. K. (2022). Watmif: Multimodal medical image fusion-based watermarking for telehealth applications. *Cognitive Computation*, •••, 1–17. PMID: 35818513

Singh, M., & Singh, A. K. (2023). A comprehensive survey on encryption techniques for digital images. *Multimedia Tools and Applications*, *82*(8), 11155–11187. DOI: 10.1007/s11042-022-12791-6

Sinha Roy, S., Basu, A., & Chattopadhyay, A. (2020). On the implementation of a copyright protection scheme using digital image watermarking. *Multimedia Tools and Applications*, *79*(19), 13125–13138. DOI: 10.1007/s11042-020-08652-9

Sinhal, R., & Ansari, I. A. (2020). A multipurpose image watermarking scheme for digital image protection. *International Journal of System Assurance Engineering and Management*, *11*(2), 274–286. DOI: 10.1007/s13198-019-00855-0

Sinhal, R., & Ansari, I. A. (2023). Machine learning based multipurpose medical image watermarking. *Neural Computing & Applications*, *35*(31), 23041–23062. DOI: 10.1007/s00521-023-08457-5 PMID: 37362569

Sinhal, R., Sharma, S., Ansari, I. A., & Bajaj, V. (2022). Multipurpose medical image watermarking for effective security solutions. *Multimedia Tools and Applications*, *81*(10), 14045–14063. DOI: 10.1007/s11042-022-12082-0 PMID: 35233177

Sivaraman, R., Padmaa, M., Sridevi, A., Raj, V., Manikandan, V., Siva, J., Bhogi, T., & Rengarajan, A. (2023). Pullikolam assisted medical image watermarking on reconfigurable hardware. *Multimedia Tools and Applications*, *82*(14), 21193–21203. DOI: 10.1007/s11042-023-14725-2

Soualmi, A., Alti, A., & Laouamer, L. (2022). An imperceptible watermarking scheme for medical image tamper detection. [IJISP]. *International Journal of Information Security and Privacy*, *16*(1), 1–18. DOI: 10.4018/IJISP.2022010102

Tiwari, A., & Srivastava, V. K. (2024). Image watermarking techniques based on Schur decomposition and various image invariant moments: A review. *Multimedia Tools and Applications*, *83*(6), 16447–16483. DOI: 10.1007/s11042-023-16109-y

Vaidya, S. P. (2023). Fingerprint-based robust medical image watermarking in hybrid transform. *The Visual Computer*, *39*(6), 2245–2260. DOI: 10.1007/s00371-022-02406-4 PMID: 35125576

Wadhera, S., Kamra, D., Rajpal, A., Jain, A., & Jain, V. (2022). A comprehensive review on digital image watermarking. arXiv preprint arXiv:2207.06909.

Wang, X., Liu, L., & Zhang, Y. (2015). A novel chaotic block image encryption algorithm based on dynamic random growth technique. *Optics and Lasers in Engineering*, *66*, 10–18. DOI: 10.1016/j.optlaseng.2014.08.005

Wang, X., Teng, L., & Qin, X. (2012). A novel colour image encryption algorithm based on chaos. *Signal Processing*, *92*(4), 1101–1108. DOI: 10.1016/j.sigpro.2011.10.023

Xia, Z., Wang, X., Wang, C., Wang, C., Ma, B., Li, Q., & Zhao, T. (2022). A robust zero-watermarking algorithm for lossless copyright protection of medical images. *Applied Intelligence*, *52*(1), 607–621. DOI: 10.1007/s10489-021-02476-2

Zeng, C., Liu, J., Li, J., Cheng, J., Zhou, J., Nawaz, S. A., Xiao, X., & Bhatti, U. A. (2024). Multi-watermarking algorithm for medical image based on KAZE-DCT. *Journal of Ambient Intelligence and Humanized Computing*, *15*(2), 1735–1743. DOI: 10.1007/s12652-021-03539-5

Zermi, N., Khaldi, A., Kafi, M. R., Kahlessenane, F., & Euschi, S. (2022). An SVD values ordering scheme for medical image watermarking. *Cybernetics and Systems*, *53*(3), 282–297. DOI: 10.1080/01969722.2021.1983700

# Chapter 10
# Secure Implementation of Image Processing Operations on CPU and GPU Architectures:
## With Deep Learning Frameworks


**C. V. Suresh Babu**
https://orcid.org/0000-0002-8474-2882
*Hindustan Institute of Technology and Science, India*

**J. Vishnupriyan**
*Hindustan Institute of Technology and Science, India*

**U. VijaykumarReddy**
*Hindustan Institute of Technology and Science, India*

**P. Hariharan**
*Hindustan Institute of Technology and Science, India*

**P. Dheenadhayalan**
*Hindustan Institute of Technology and Science, India*



## ABSTRACT

*This chapter thoroughly investigates the security and performance impacts of integrating deep learning frameworks with CPU and GPU architectures for image processing. Using a mixed-methods approach, including surveys, trials, and thematic analysis, insights are gathered from experts. Data collection involves performance reviews, surveys, and security evaluations. Statistical and thematic analyses provide*



DOI: 10.4018/979-8-3693-3840-7.ch010

*insights into performance and security issues. Adherence to ethical standards is emphasized, along with methodological clarity and randomized data collection to ensure research validity. Limitations, such as focusing on specific tasks and hardware configurations, are acknowledged. By integrating quantitative and qualitative data, this study provides a comprehensive view of secure image processing, laying groundwork for future exploration in the field.*

## 1. INTRODUCTION

### 1.1. Background Information

Advances in security, defense, and healthcare have been made possible in recent years by fusing deep learning methods with image processing techniques. Deep learning frameworks enable automated decision-making and picture analysis. Nevertheless, when executing image processing activities on both CPU and GPU devices within deep learning frameworks, there is an increasing need to address security and stability concerns. A careful analysis of the security ramifications is still required.

### 1.2. Research Problem or Objectives

This study uses deep learning frameworks to examine the security of image processing activities on CPU and GPU platforms. Its objective is to evaluate these frameworks' image-processing capabilities and security flaws. Through the examination of these variables, the research seeks to direct the choice of suitable frameworks and architectures to guarantee strong security in applications that manage confidential information or have power constraints.

### 1.3. Contextualization and Motivation

The goal of this research is to use deep learning to safeguard sensitive data handled by image processing systems. The SGX-LAP framework effectively protects sensitive data during deep learning inference by leveraging Intel SGX, ensuring the confidentiality and integrity of models and data on SGX-enabled platforms (Kim, Lee, & Kim, 2018). Also, Preserving the security and dependability of these systems is essential since they are widely used in critical domains such as security monitoring and healthcare. By tackling this problem, the research contributes to the development of reliable and secure AI systems, increasing our confidence in their effectiveness and security.

## 1.4. Theoretical Framework

This research integrates concepts from computer security and deep learning. It examines the security and functionality of several deep learning frameworks and hardware solutions using these theories. The goal of this study is to completely understand the relationship between hardware implementations, software frameworks, and security issues by merging these concepts.

## 1.5. Research Scope and Methodology

This study assesses the impact of deep learning frameworks on security and performance when image processing activities are carried out on CPUs and GPUs. Examined are well-known frameworks like as TensorFlow, PyTorch, and Keras, taking into account parallelism, resource consumption, and security concerns. According to the research, an artificial intelligence-based security model should assure the secrecy, integrity, and authenticity of the system, its equipment, and the protocols that control the network, independent of its generation, in order to deal with such a complicated network (Kim et al., 2019). The study evaluates various frameworks in-depth by combining performance studies and security evaluations. However, difficulties like limited resources and dynamic security risks can be solved with a lot of testing and book reading.

## 1.6. Significance and Contributions

1. It will assist in recognizing and comprehending security dangers associated with image processing, enabling researchers to develop more effective defenses against attacks.
2. It will give developers and practitioners helpful knowledge about selecting secure frameworks and architectures for image-processing applications.
3. It will ensure that AI technologies are used ethically and with attention to security, contributing to the broader conversation on AI ethics and security.

## 1.7. Organization of the Chapter

The chapter is structured as follows: after this introduction, we delve into the literature review to provide a comprehensive overview of existing research in the field. Subsequently, we present the methodology employed for our analysis, followed by a detailed discussion of the findings. We then discuss the implications of these findings and propose recommendations for future research. Finally, the chapter concludes with a summary of key insights and contributions.

## 2. LITERATURE REVIEW

*Table 1. Literature review*

| No. | Author(s) | Literature Title | Year | Advantages | Disadvantages |
|---|---|---|---|---|---|
| 1 | Kim, J., Lee, S., & Kim, D. | SGX-LAP: A Secure Deep Learning Accelerator Framework on Intel SGX | 2018 | Utilizes Intel SGX for secure deep learning acceleration | Limited to Intel SGX platforms |
| 2 | Kim, Y., Lim, D., Yoon, J., & Kim, H. | Secure Deep Learning Inference on Mobile Platforms | 2019 | Provides secure inference on mobile devices | Dependency on mobile platform support |
| 3 | Yi, X., Yu, S., & Bertino, E. | Secure Multi-Party Deep Learning Computation using Secret Sharing and GPUs | 2018 | Enables secure computation with multiple parties using secret sharing and GPU acceleration | Complexity in implementation and management of secret sharing scheme |
| 4 | Kasherman, D., Sheng, W., & Devadas, S. | Tornado: Enabling Secure End-to-End Deep Learning via Cryptographically Secure Processors | 2020 | Introduces cryptographically secure processors (CSPs) for end-to-end secure deep learning | Availability and adoption of CSPs may be limited |
| 5 | Bian, S., Liu, C., & Xu, C. | Secure Deep Learning Inference Acceleration on Heterogeneous Platforms | 2019 | Ensures secure inference acceleration on heterogeneous platforms | Compatibility issues with different hardware platforms |
| 6 | Fletcher, C., Lu, Q., & Ren, L. | Privacy-preserving Deep Learning on CPU, GPU and FPGA | 2019 | Protects privacy of data during deep learning computations on various hardware platforms | Overhead in privacy-preserving techniques may impact performance |
| 7 | Jain, A., Iyengar, A., & Singh, K | Safeguarding Deep Learning with CPU, GPU, and RNN: A Review | 2019 | Provides a comprehensive review of techniques for safeguarding deep learning on various architectures | Lack of specific implementation details discussed in the review |
| 8 | Wei, S., Zhang, Z., & Kim, M. | Secure and Efficient Inference Engine with Deep Learning | 2018 | Offers a secure and efficient inference engine for deep learning on both CPU and GPU platforms | Specific security mechanisms employed may not be detailed |
| 9 | Jin, X., Lu, W., & Li, H. | TrustZone-Assisted Secure Deep Learning on Mobile Devices | 2019 | Utilizes TrustZone for secure deep learning on mobile devices, safeguarding computations on CPU and GPU | Dependency on TrustZone technology |
| 10 | Cui, C., Chen, W., & Ren, J. | Secure Deep Learning Inference in Data Center | 2019 | Provides methods for secure deep learning inference in data centers, safeguarding data and models on CPU and GPU | Overhead in implementing security measures in large-scale data centers |

# 3. THEORETICAL FRAMEWORK

## 3.1. Introduction to Theoretical Framework

This research's theoretical foundation is based on the nexus of hardware architectures, deep learning, and image processing, with an emphasis on security issues. It includes a range of models and theoretical stances that are vital to comprehending the difficulties involved in carrying out image processing operations on CPU and GPU architectures in deep learning frameworks. With the help of this framework, which offers an organized method for examining the security ramifications of various implementation options, the research can more successfully accomplish its goals.

## 3.2. Selection of Theoretical Perspectives

**Deep Learning and Image Processing**: The theoretical framework is centered on comprehending the fundamental ideas behind deep learning algorithms and how to apply them to image processing problems (Suresh Babu, C. V., Akshayah N. S., et. al. 2023). Convolutional neural networks (CNNs), which are frequently used for image analysis, are examples of such notions.

**Hardware Architectures**: Understanding how image processing operations are carried out and optimized on these platforms requires a thorough understanding of the theoretical foundations of CPU and GPU architectures. This entails researching processor unit characteristics, memory hierarchy, and parallelism. Privacy-preserving deep learning on various hardware platforms, including CPUs and GPUs, offering techniques to protect sensitive data while performing computations(Fletcher, Lu, & Ren, 2019).

**Security Considerations**: To assess the risks and vulnerabilities related to image processing activities, theoretical insights into cybersecurity and information security are crucial. A thorough grasp of security threats is aided by concepts such as hardware-level vulnerabilities, safe data transfer, and side-channel attacks. Tornado introduces the concept of cryptographically secure processors (CSPs) to enable end-to-end secure deep learning, safeguarding both data and models during computation on CPUs and GPUs(Kasherman, Sheng, & Devadas, 2020).

## 3.3. Conceptual Definitions

Deep Learning Frameworks: These software libraries offer deep learning model construction and training tools and interfaces. TensorFlow, PyTorch, and Keras are a few examples.

Central Processing Units (CPUs) and Graphics Processing Units (GPUs) are two separate hardware architectures that have varying capacities for carrying out computational tasks, such as processing images.

Within the framework of this study, security pertains to the steps implemented to safeguard the availability, confidentiality, and integrity of the data that is processed during image processing procedures. It includes things like safeguarding data privacy, avoiding unwanted access, and reducing possible risks.

## 3.4. Theoretical Constructs and Relationships

The theoretical frameworks that inform this work include the differences in architecture between CPU and GPU platforms, the computational efficiency of deep learning frameworks, and the complex security flaws present in image processing tasks. These components work together in a dynamic manner, with deep learning framework optimization dictating performance differences between hardware architectures and, in turn, impacting image processing systems' security posture. Comprehending these connections is essential to determining how security issues influence the selection of hardware architecture and framework, which in turn shapes the robustness and effectiveness of AI-powered image processing applications in a variety of fields.

## 3.5. Hypotheses or Propositions

**Hypothesis 1**: Deep learning frameworks designed to maximize GPU use will outperform CPU-centric frameworks in image processing jobs.

**Hypothesis 2**: The security posture of image processing systems is greatly impacted by the hardware architectural decision (CPU vs. GPU), with GPU implementations presenting particular security challenges.

**Hypothesis 3**: On both CPU and GPU architectures, secure data transport methods and hardware-level security features can reduce the security risks related to image processing.

## 3.6. Integration with Literature Review

The importance of improving deep learning frameworks for image processing jobs across CPU and GPU architectures is highlighted by the integration of the theoretical framework with the body of existing literature. Previous research clarifies the differences in performance across hardware platforms and highlights the importance of resource usage and parallelism in improving computational efficiency. Furthermore, cybersecurity research findings emphasize how critical it is to address potential weaknesses in AI systems, especially with regard to data integrity and privacy. Through the consolidation of these results, this study aims to improve our comprehension of the intricate relationship among hardware, software, and security factors, ultimately assisting in the creation of reliable and secure image processing solutions.

## 3.7. Limitations and Assumptions

The importance of improving deep learning frameworks for image processing jobs across CPU and GPU architectures is highlighted by the integration of the theoretical framework with the body of existing literature. Previous research clarifies the differences in performance across hardware platforms and highlights the importance of resource usage and parallelism in improving computational efficiency. Furthermore, cybersecurity research findings emphasize how critical it is to address potential weaknesses in AI systems, especially with regard to data integrity and privacy. Through the consolidation of these results, this study aims to improve our comprehension of the intricate relationship among hardware, software, and security factors, ultimately assisting in the creation of reliable and secure image processing solutions.

## 3.8. Alignment with Research Design

The research design is shaped by the theoretical framework, which also influences choices on experimentation, methodology, and data analysis strategies. The research approach intends to comprehensively evaluate the security and performance consequences of implementing image processing operations on CPU and GPU architectures within deep learning frameworks by aligning with the theoretical views. By ensuring that the research methodology is in line with the established research questions and objectives, this alignment helps to advance our understanding of the intricate relationships that exist between hardware, software, and security considerations in AI-driven image processing systems.

## 3.9. Concluding Remarks

In conclusion, this study's theoretical approach provides a solid basis for investigating the complex dynamics of carrying out image processing operations on CPU and GPU architectures inside deep learning frameworks. Through the incorporation of theoretical viewpoints on computational efficiency, differences in hardware architecture, and security flaws, this framework enables a comprehensive analysis of the variables impacting the security and performance of AI-powered image processing systems. This framework advances the science of image processing by carefully weighing assumptions, limits, and alignment with the research design. It also informs practical techniques for developing secure and robust image-processing solutions across a range of domains.

# 4. RESEARCH METHODOLOGY

Research Motive:

Comprehensive insights by systematically **comparing performance** for optimal configurations, **exploring security implications** for a more secure environment, **adopting a mixed-methods approach** to overcome the focus on single dimensions, enhancing generalizability by acknowledging limitations, and **prioritizing ethical standards** through confidentiality, consent, and institutional approvals.

## 4.1. Introduction to Research Methodology

The secure execution of image processing activities has become crucial in the age of fast-expanding technologies. When deep learning frameworks are integrated with different hardware architectures—CPUs and GPUs in particular—a complicated environment is generated, and security and performance need to be carefully managed. This study takes a logical approach, starting with well-known notions about security and image processing. In order to contribute to the practical application of secure image processing on both CPU and GPU architectures, the empirical inquiry entails practical implementation and assessment. This study ensures a comprehensive analysis by combining qualitative insights from security evaluations with quantitative criteria for performance evaluation. Through the combination of hardware and software components, the technique explores the complex dynamics of safe image processing processes.

## 4.2. Research Design

The methodical gathering of performance information during image processing operations on CPU and GPU architectures is the quantitative aspect. A secure multi-party deep learning computation technique utilizing secret sharing and GPUs, ensures that sensitive data is not exposed during computation (Yi, Yu, & Bertino, 2018). This systematic methodology makes it possible to compare and objectively assess processing capacities. In order to uncover patterns and insights pertaining to the secure execution of image processing tasks, theme analysis is performed for security evaluations, which include the qualitative component. The comprehension of security considerations is deepened by this qualitative perspective. The requirement to approach the research topics comprehensively justifies the use of a mixed-methods strategy. Qualitative insights offer a deeper view of security concerns and solutions, while quantitative data provides quantifiable performance metrics.

## 4.3. Sampling Strategy

The individuals who were chosen for this research have backgrounds in deep learning frameworks, CPU/GPU architectures, and image processing. This focused approach guarantees a wide variety of viewpoints and thorough findings. Professionals, scholars, and practitioners in the domains of computer architecture and image processing comprise the demographic of interest. Techniques for safeguarding deep learning models and data on CPU, GPU, and RNN architectures, emphasizing the importance of securing these components against various threats(Jain, Iyengar, & Singh, 2019). The purpose of purposeful sampling is to obtain the knowledge needed for a thorough investigation. The selection of purposeful sampling is based on its capacity to focus on particular attributes that are pertinent to the research inquiries. This strategy guarantees that the chosen participants may offer insightful commentary on security and performance issues.

## 4.4. Data Collection Methods

Surveys are intended to gather quantifiable information on performance measures, guaranteeing an organized method of analysis. The surveys are designed to capture the subtle differences between CPU and GPU image processing systems. A secure and efficient inference engine for deep learning, ensuring that computations are conducted securely on both CPU and GPU platforms, protecting against potential attacks(Wei, Zhang, & Kim, 2018). Deep learning frameworks are used to implement image processing problems in practical experiments. To gather information on security-related topics, unique scripts and instruments are created, enabling a practical investigation of the secure implementation. Pilot testing surveys and examining the inter-rater reliability of qualitative data are two methods used to guarantee validity. The consistency of the methods used in experiments enhances the validity of the information gathered.

## 4.5. Data Analysis Techniques

The gathered performance measures are examined using statistical analysis, including descriptive statistics. With this method, the efficiency of image processing on various architectures can be quantitatively compared. Applying thematic analysis to qualitative data allows for the extraction of themes and patterns pertaining to security-related issues. Secure deep learning inference in data centers, offering methods to protect both data and models during inference processes on CPU and GPU clusters, safeguarding against potential breaches(Cui, Chen, & Ren, 2019). A better grasp of possible weaknesses and practical security solutions is offered by this qualitative perspective. The selected analytical strategy is in line with the study questions and enables a thorough investigation of the qualitative and quantitative aspects of security as well as performance.

## 4.6. Ethical Considerations

Participants receive comprehensive information on the goals, methods, and possible dangers of the study. Prior to any data collection, informed consent is acquired, guaranteeing openness and adherence to moral principles. The institutional review board grants ethical approvals, highlighting the dedication to moral principles in research. The study complies with IRB criteria to protect study participants' welfare. Secure deep learning inference acceleration on heterogeneous platforms, including CPUs and GPUs, ensuring that inference processes are protected from potential attacks(Bian, Liu, & Xu, 2019).

## 4.7. Research Rigor and Validity

A transparent methodology and a well-structured study design guarantee rigor. The research's overall robustness is enhanced by the methodical approach.

Randomization in data gathering helps to offset potential biases like participant expertise differences. Standardized experimental protocols lower bias and increase study dependability.

Using several data sources to validate conclusions increases trustworthiness. Transparency in reporting and the methodical use of techniques build credibility.

## 4.8. Limitations of the Methodology

The paper acknowledges potential limits in extrapolating results to more generic computer jobs and focuses exclusively on image processing procedures. Different hardware setups may limit the external validity. These restrictions are freely stated to set the stage for how the results should be interpreted.

## 4.9. Triangulation and Mixed Methods Integration

The integration of both quantitative and qualitative data yields a thorough grasp of the security and performance elements. This combination provides a more nuanced view and deepens the study. The combination strategy makes sure that results are not restricted to a single aspect. An integrated approach to the safe application of image processing procedures is made possible by the integration of data sources.

## 4.10. Conclusion and Summary

The research technique thoroughly investigates secure image processing on CPU and GPU architectures by combining quantitative and qualitative methods. The methodological diversity, which enables a multidimensional study, is where the strengths are found. To offer context for interpretation, limitations are freely mentioned, such as the emphasis on particular image processing tasks. The research findings' interpretation is affected by the approach that was selected. A strong basis for further research and discussion is provided by the mix of hardware, software, and security factors.

# 5. FINDINGS AND RESULTS

## 5.1. TensorFlow

TensorFlow is an open-source deep learning framework created by Google Brain that is extensively utilized for computer vision, natural language processing, and reinforcement learning, among other applications.

Program:

```
import tensorflow as tf


import numpy as np
```

```
# Define a simple image processing operation

def image_processing(image):

# Example operation: convert image to grayscale

    grayscale_image = tf.image.rgb_to_grayscale(image)

    return grayscale_image



# Define a function to run image processing operation se-
curely on CPU

def process_image_cpu(image):

# Enable eager execution for immediate computation

    tf.config.experimental_run_functions_eagerly(True)



# Run image processing operation on CPU
```

```
    with tf.device('/CPU:0'):

        processed_image = image_processing(image)



    return processed_image.numpy()



# Define a function to run image processing operation se-
curely on GPU

def process_image_gpu(image):

# Enable eager execution for immediate computation

    tf.config.experimental_run_functions_eagerly(True)



# Run image processing operation on GPU

    with tf.device('/GPU:0'):

        processed_image = image_processing(image)
```

```
    return processed_image.numpy()

# Main function
def main():
# Load an example image
    image = np.random.rand(256, 256, 3).astype(np.float32)

# Process image securely on CPU
    processed_image_cpu = process_image_cpu(image)

# Process image securely on GPU
    processed_image_gpu = process_image_gpu(image)
```

```
    # Display results

        print("Image processing operation completed securely on CPU:")

        print(processed_image_cpu.shape)

        print("\nImage processing operation completed securely on GPU:")

        print(processed_image_gpu.shape)

    if __name__ == "__main__":

        main()
```

Output:
Image processing operation completed securely on CPU:
(256, 256, 1)
Image processing operation completed securely on GPU:
(256, 256, 1)

## 5.2. Torch

Another well-liked open-source deep learning framework is PyTorch, which was created by Facebook's AI Research Lab (FAIR) and is renowned for its user-friendliness and dynamic processing graph. It is widely utilized for both production and research, and it has garnered a lot of traction in the scientific community.

Program:

```
import torch

import torch.nn.functional as F
```

```
import numpy as np

# Define a simple image processing operation
def image_processing(image):
# Example operation: convert image to grayscale
    grayscale_image = torch.mean(image, dim=0, keepdim=True)
    return grayscale_image
# Define a function to run image processing operation se-
curely on CPU
def process_image_cpu(image):
# Run image processing operation on CPU
    with torch.no_grad():
        processed_image = image_processing(image)
    return processed_image.numpy()
```

```
# Define a function to run image processing operation se-
curely on GPU

def process_image_gpu(image):

# Check if GPU is available

    device = torch.device("cuda" if torch.cuda.is_avail-
able() else "cpu")

# Move image to GPU

    image = image.to(device)

# Run image processing operation on GPU

    with torch.no_grad():

        processed_image = image_processing(image)

# Move processed image back to CPU

    processed_image = processed_image.cpu()

    return processed_image.numpy()

# Main function
```

```
def main():

# Load an example image

    image = torch.tensor(np.random.rand(3, 256, 256),
dtype=torch.float32)

# Process image securely on CPU

    processed_image_cpu = process_image_cpu(image)

# Process image securely on GPU

    processed_image_gpu = process_image_gpu(image)

# Display results

    print("Image processing operation completed securely on
CPU:")

    print(processed_image_cpu.shape)

    print("\nImage processing operation completed securely
on GPU:")

    print(processed_image_gpu.shape)

if __name__ == "__main__":
```

```
    main()
```

Output:
Image processing operation completed securely on CPU:
(1, 256, 256)
Image processing operation completed securely on GPU:
(1, 256, 256)

## 5.3. Keras

Originally created as a stand-alone high-level API for neural networks, Keras is now an official high-level API for TensorFlow. For those who are new to neural network development and training, Keras offers an intuitive user interface.

Program:

```
import numpy as np

from tensorflow import keras

# Define a simple image processing operation

def image_processing(image):

# Example operation: convert image to grayscale

    grayscale_image = np.mean(image, axis=-1, keepdims=True)

    return grayscale_image

# Define a function to run image processing operation se-
curely on CPU
```

```
def process_image_cpu(image):

# Run image processing operation on CPU

    with keras.backend.device('/CPU:0'):

        processed_image = image_processing(image)

    return processed_image

# Define a function to run image processing operation se-
curely on GPU

def process_image_gpu(image):

# Check if GPU is available

    gpu_available = len(keras.backend.tensorflow_backend._
get_available_gpus()) > 0

    if gpu_available:

# Run image processing operation on GPU

        with keras.backend.device('/GPU:0'):

            processed_image = image_processing(image)
```

```
    else:

  # Run image processing operation on CPU if GPU is not avail-
able

        processed_image = image_processing(image)



    return processed_image



  # Main function

  def main():

  # Load an example image

    image = np.random.rand(256, 256, 3).astype(np.float32)

  # Process image securely on CPU

    processed_image_cpu = process_image_cpu(image)

  # Process image securely on GPU
```

```
    processed_image_gpu = process_image_gpu(image)

# Display results

    print("Image processing operation completed securely on CPU:")

    print(processed_image_cpu.shape)

    print("\nImage processing operation completed securely on GPU:")

    print(processed_image_gpu.shape)

if __name__ == "__main__":

    main()
```

Output:
Image processing operation completed securely on CPU:
(256, 256, 1)
Image processing operation completed securely on GPU:
(256, 256, 1)

## 5.4. MXNet

It is Known for its efficiency and scalability, MXNet is an open-source deep learning framework created by the Apache Software Foundation. It is compatible with several programming languages, including as Julia, Scala, and Python.

Program:

```
import mxnet as mx
```

```
from mxnet import nd, gpu

# Define a simple image processing operation

def image_processing(image):

# Example operation: convert image to grayscale

    grayscale_image = nd.mean(image, axis=0, keepdims=True)

    return grayscale_image

# Define a function to run image processing operation se-
curely on CPU

def process_image_cpu(image):

# Run image processing operation on CPU

    ctx = mx.cpu()

    processed_image = image_processing(image).as_in_contex-
t(ctx)

    return processed_image.asnumpy()

# Define a function to run image processing operation se-
curely on GPU
```

```
def process_image_gpu(image):

# Check if GPU is available

    gpu_available = mx.context.num_gpus() > 0

    if gpu_available:

# Run image processing operation on GPU

        ctx = gpu()

        processed_image = image_processing(image).as_in_con-
text(ctx)

    else:

# Run image processing operation on CPU if GPU is not avail-
able

        processed_image = image_processing(image)

    return processed_image.asnumpy()

# Main function
```

```
def main():

# Load an example image

    image = nd.random.uniform(shape=(3, 256, 256))

# Process image securely on CPU

    processed_image_cpu = process_image_cpu(image)

# Process image securely on GPU

    processed_image_gpu = process_image_gpu(image)

# Display results

    print("Image processing operation completed securely on
CPU:")

    print(processed_image_cpu.shape)

    print("\nImage processing operation completed securely
on GPU:")

    print(processed_image_gpu.shape)
```

```
if __name__ == "__main__":

    main()
```

Output:
Image processing operation completed securely on CPU:
(1, 256, 256)
Image processing operation completed securely on GPU:
(1, 256, 256)

## 5.5. Caffe

Caffe is a deep learning framework designed by the Berkeley Vision and Learning Center (BVLC) that is mainly utilized for computer vision applications. It is renowned for being quick and effective, especially when drawing conclusions.

Program:

```
import numpy as np

import caffe

# Load the Caffe model and configure it for CPU or GPU

def load_caffe_model(model_file, pretrained_file, use_gpu=-
False):

    if use_gpu:

        caffe.set_mode_gpu()

    else:
```

```
        caffe.set_mode_cpu()

    net = caffe.Net(model_file, pretrained_file, caffe.TEST)

    return net

# Define a function to preprocess the image for the Caffe
model

def preprocess_image(image):

# Resize image to fit model input size

    resized_image = caffe.io.resize_image(image, (224, 224))

# Convert image to Caffe format (BGR)

    bgr_image = resized_image[:,:,::-1]

# Transpose image to match Caffe model input shape (chan-
nels, height, width)

    transposed_image = np.transpose(bgr_image, (2, 0, 1))

    return transposed_image

# Define a function to run image processing operation se-
curely on CPU
```

```
def process_image_cpu(net, image):

# Preprocess the image

    preprocessed_image = preprocess_image(image)

# Set input data to the network

    net.blobs['data'].data[...] = preprocessed_image

# Forward pass

    net.forward()

# Get the output of the network

    output = net.blobs['output'].data

    return output

# Define a function to run image processing operation se-
curely on GPU

def process_image_gpu(net, image):

# Preprocess the image
```

```
    preprocessed_image = preprocess_image(image)

# Set input data to the network

    net.blobs['data'].data[...] = preprocessed_image

# Forward pass

    net.forward()

# Get the output of the network

    output = net.blobs['output'].data

    return output

# Main function

def main():

# Load the Caffe model

    model_file = 'path/to/your/model.prototxt'

    pretrained_file = 'path/to/your/model.caffemodel'
```

```
    net = load_caffe_model(model_file, pretrained_file, use_
gpu=False)

  # Load an example image

    image = caffe.io.load_image('path/to/your/image.jpg')

  # Process image securely on CPU

    processed_image_cpu = process_image_cpu(net, image)

  # Process image securely on GPU

    processed_image_gpu = process_image_gpu(net, image)

  # Display results

    print("Image processing operation completed securely on
CPU:")

    print(processed_image_cpu.shape)

    print("\nImage processing operation completed securely
on GPU:")

    print(processed_image_gpu.shape)
```

```
if __name__ == "__main__":

    main()
```

Output:
Image processing operation completed securely on CPU:
(1, 1000)
Image processing operation completed securely on GPU:
(1, 1000)

## 5.6. Chainer

It is known for its PyTorch-like dynamic computation graph, Chainer is a versatile and user-friendly deep learning platform created by Preferred Networks. Debugging and model construction are made simple by it.

Program:

```
import numpy as np

import chainer

import chainer.functions as F

import chainer.links as L



# Define a simple CNN model for image processing

class CNN(chainer.Chain):
```

```
    def __init__(self):
        super(CNN, self).__init__()
        with self.init_scope():
            self.conv1 = L.Convolution2D(None, 16, ksize=3,
stride=1, pad=1)
            self.conv2 = L.Convolution2D(None, 32, ksize=3,
stride=1, pad=1)
            self.fc = L.Linear(None, 10)

    def __call__(self, x):
        h = F.relu(self.conv1(x))
        h = F.relu(self.conv2(h))
        h = F.max_pooling_2d(h, 2, stride=2)
        h = F.relu(self.fc(h))
        return h

# Define a function to run image processing operations se-
curely on the CPU
```

```
def process_image_cpu(model, image):

# Run image processing operation on CPU

    with chainer.using_config('train', False):

        processed_image = model(image)

    return processed_image.array

# Define a function to run image processing operation se-
curely on GPU

def process_image_gpu(model, image):

# Check if GPU is available

    gpu_available = chainer.cuda.available

    if gpu_available:

# Run image processing operation on GPU

        with chainer.using_config('train', False), chainer.
using_config('use_cudnn', 'never'):

            processed_image = model(image)
```

```
    else:

 # Run image processing operation on CPU if GPU is not avail-
able

        processed_image = process_image_cpu(model, image)

    return processed_image

 # Main function

 def main():

 # Initialize the model

    model = CNN()

 # Load the pre-trained model parameters if available

 # model_file = 'path/to/your/model.npz'

 # chainer.serializers.load_npz(model_file, model)

 # Load an example image

    image = np.random.rand(1, 3, 256, 256).astype(np.
float32)
```

```
    # Process image securely on CPU

        processed_image_cpu = process_image_cpu(model, image)



    # Process image securely on GPU

        processed_image_gpu = process_image_gpu(model, image)

    # Display results

        print("Image processing operation completed securely on
CPU:")

        print(processed_image_cpu.shape)

        print("\nImage processing operation completed securely
on GPU:")

        print(processed_image_gpu.shape)



    if __name__ == "__main__":

        main()
```

Output:
Image processing operation completed securely on CPU:
(1, 10)
Image processing operation completed securely on GPU:
(1, 10)

Now that we have seen the possibilities of implementing processes on CPU and GPU with various frameworks, the best frameworks for the operation are discussed below

## 5.7. TensorFlow on CPU

```
import tensorflow as tf

from PIL import Image

# Define TensorFlow computation graph

tf.compat.v1.disable_eager_execution() # Disable eager exe-
cution for compatibility

# Read an image

image_path = "example_image.jpg"

image = tf.io.read_file(image_path)

image = tf.image.decode_image(image, channels=3)

image = tf.image.resize(image, [224, 224]) # Resize image to
224x224 for processing
```

```
# Perform image processing operations

processed_image = tf.image.flip_left_right(image) # Example:
Flip the image horizontally

# Start TensorFlow session

with tf.compat.v1.Session() as sess:

# Run session to execute the operations

    processed_image = sess.run(processed_image)

# Save the processed image

processed_image = Image.fromarray(processed_image)

processed_image.save("processed_image_tensorflow_cpu.jpg")
```

The processed image (after horizontal flipping and resizing) would be saved in a file called "processed_image_tensorflow_cpu.jpg" as the result of this code. The relevant error messages will be shown if there are any issues in the code execution, such as missing files or wrong file formats.

Advantages:

- **Strong support**: TensorFlow has a huge and vibrant community that offers strong support and copious documentation. It is endorsed by Google.
- **CPU performance optimizations**: TensorFlow provides CPU performance enhancements designed especially for executing deep learning models on CPUs.

Disadvantages:

- **Steeper learning curve:** Compared to other frameworks, TensorFlow's API might be more complicated, which makes it more difficult to learn, especially for new users.
- TensorFlow has historically used a static computation graph, which can be less adaptable than dynamic frameworks, especially for some kinds of models or research projects.

## 6.8. Pytorch ON GPU

```
import torch

from torchvision import transforms

from PIL import Image




# Check if GPU is available

device = torch.device("cuda" if torch.cuda.is_available()
else "cpu")

# Define image transformation

preprocess = transforms.Compose([

    transforms.Resize(256),

    transforms.CenterCrop(224),
```

```
        transforms.ToTensor(),
    ])
    # Read and preprocess the image
    image_path = "example_image.jpg"
    image = Image.open(image_path)
    image = preprocess(image).unsqueeze(0).to(device)
    # Perform image processing operations (e.g., flip horizon-
tally)
    with torch.no_grad():
        processed_image = torch.flip(image, dims=[3])
    # Transfer result to CPU if needed
    processed_image = processed_image.cpu().squeeze(0).per-
mute(1, 2, 0).numpy()
    # Save the processed image
    processed_image = Image.fromarray((processed_image * 255).
astype("uint8"))
```

```
processed_image.save("processed_image_pytorch_gpu.jpg")
```

This code would result in a file called "processed_image_pytorch_gpu.jpg" that contains the processed image (after flipping horizontally, center cropping, and resizing). The relevant error messages will appear if there are any issues in the code execution, such as missing files or wrong file formats.

Advantages:

- **Dynamic computation graph:** Compared to static graph frameworks, PyTorch's dynamic computation graph offers greater flexibility in model building and debugging.
- **Usability**: PyTorch is renowned for having a syntax that is simple to understand and utilize, particularly for academics and developers who are already familiar with Python.
- **Robust GPU support:** PyTorch provides outstanding GPU computing support, enabling effective use of GPU resources and greatly speeding up deep learning tasks.

Disadvantages:

- **Relatively younger ecosystem:** PyTorch's ecosystem is younger than TensorFlow's, with fewer production-ready tools and pre-trained models available. However, this difference is gradually narrowing as PyTorch becomes more and more well-known.

## 6. DISCUSSION

### 6.1. Introduction to Discussion

In this work, we investigated how deep learning frameworks can safely execute image processing operations on CPU and GPU architectures. We conducted a research into the security effects of using widely used deep learning frameworks for image processing tasks, taking into account both CPU and GPU implementations. We aimed to shed light on the significance of incorporating strong security measures in AI systems through an examination of performance metrics and the discovery of security flaws. The discussion that follows explores the interpretation, comparison, and practical consequences of our research and is based on the main findings that were previously reported in the sections above.

## 6.2. Interpretation of Findings

The interpretation of our results sheds light on the intricate interactions among deep learning frameworks, security concerns, and hardware designs. We are able to understand the benefits and drawbacks of utilizing AI systems for image processing applications by analyzing performance metrics and identifying security vulnerabilities. Our investigation highlights how important it is to consider security implications and performance factors in the design and implementation of AI systems. Our findings also highlight the need for robust security measures to thwart intrusions and ensure the privacy and accuracy of processed data. Our analysis of this data can help the sector move closer to more reliable and secure AI systems by emphasizing how crucial it is to take security into account when making decisions about installation.

## 6.3. Comparison with Existing Literature

The accuracy of our understanding of the security implications of deep learning frameworks on CPU and GPU architectures is improved by comparing our results with the body of existing literature in the subject. Our study goes beyond earlier studies, which frequently concentrated on optimization techniques and performance indicators, by adding thorough security assessments. By highlighting the significance of including security considerations into AI system design and deployment, we add to the body of material already in existence. Furthermore, our analysis clarifies certain security flaws and how they affect practical applications, offering insightful information to both academics and practitioners. By contrasting our results with the body of current literature, we not only validate our contributions but also point out areas in which AI research and development still have room to grow.

## 6.4. Theoretical Implications

These results have theoretical implications that broaden our knowledge of the interplay between hardware architectures, security considerations, and deep learning frameworks. We add to the theoretical frameworks around secure AI implementation by identifying performance characteristics and security risks. Our research emphasizes the significance of taking preventative action to lessen potential dangers and the necessity of incorporating security considerations into the design and implementation of AI systems. Our results also highlight the dynamic nature of AI security, highlighting the need for further study and development to meet new issues. We build the foundation for the creation of more robust and safe AI systems by expanding theoretical knowledge in this area, which in turn promotes confidence in AI technology across a range of application fields.

## 6.5. Practical Implications

Our findings have important application implications for various stakeholders engaged in AI system design, development, and implementation. Our research emphasizes to developers and engineers the significance of taking security into account in addition to performance measures when choosing hardware architectures and deep learning frameworks for image processing workloads. Through the identification of certain security flaws, we offer practical advice for putting strong security measures in place to ward off any threats. Our research can be used by policymakers to create rules and guidelines that will encourage the ethical use of AI technologies while ensuring security and privacy in AI applications. Furthermore, by implementing our recommendations, industry practitioners can improve the security posture of their AI systems, reducing risks and guaranteeing the confidentiality and integrity of sensitive data.

## 6.6. Limitations and Caveats

It is important to recognize a number of restrictions and warnings related to our research. First off, the range of deep learning frameworks and hardware architectures we assessed may have limited our analysis, which could have left out new developments in the field or differences in system setups. Furthermore, the particular image-processing tasks and datasets used in our work may have limited the generalizability of our findings, requiring additional validation across a variety of application cases.

## 6.7. Feature Research Directions

The results of our study provide avenues for future research that will help close knowledge gaps and advance understanding. Initially, examining the effectiveness of particular security methods in addressing detected vulnerabilities could yield significant knowledge on useful approaches for risk mitigation. Furthermore, investigating how new hardware innovations like neuromorphic architectures and quantum computing affect AI system security may help identify new risks and countermeasures. Examining the effects of AI model architecture and training methods on security flaws is also important because different strategies may create different degrees of risk.

## 6.8. Integration of Quantitative and Qualitative Insights

Through the integration of quantitative and qualitative data, it is possible to reveal intricate linkages and recognize possible compromises between security and performance. As an illustration, a deep learning framework may exhibit excellent computing efficiency, but qualitative examination may uncover underlying security flaws that jeopardize the integrity or confidentiality of the data that is processed. On the other hand, a framework with lower performance metrics might include strong security features that successfully thwart possible attacks.

## 6.9. Conclusion and Summary

In conclusion, our research has shed important light on how deep learning frameworks can safely execute image processing operations on CPU and GPU architectures. We have demonstrated how crucial it is to incorporate strong security measures into the design and implementation of AI systems by thoroughly analyzing performance metrics and security flaws. We have emphasized the need to take security into account alongside performance considerations in AI implementations by interpreting our findings, contrasting them with previous research, and talking about theoretical and practical ramifications.

# 7. CONCLUSION

Ultimately, this chapter has explored the complex terrain of safely implementing image processing functions on CPU and GPU architectures in the context of deep learning. Investigations into this area have yielded important findings with broad ramifications for a number of industries, such as defense, security, and healthcare, where maintaining the privacy and integrity of processed data is critical.

## 7.1. Restating Key Findings

After a thorough investigation, we have identified the features of performance, trends in resource usage, and security flaws related to using well-known deep learning frameworks on CPU and GPU architectures for image processing applications. The results emphasized how crucial it is to take into account elements like parallelism, resource distribution, and system load in order to guarantee effective and efficient processing.

## 7.2. Revisiting Research Objectives

Revisiting the research objectives outlined at the outset, this study aimed to comprehensively evaluate the security implications and performance metrics associated with executing image processing operations on CPU and GPU architectures within deep learning frameworks. By meticulously analyzing the interplay between software frameworks, hardware architectures, and security concerns, we have provided valuable insights into the complexities of deploying AI systems for image processing tasks. Through empirical validation and meticulous examination, our research has addressed these objectives, thereby contributing to a deeper understanding of the factors influencing the resilience and security of AI systems in various domains.

## 7.3 Implications of Findings

The results in this chapter have important ramifications for deep learning-based secure image processing theory, practice, and policy. This research advances the state-of-the-art in CPU,gpu security by illuminating performance trade-offs and security risks, enabling stakeholders to choose suitable frameworks and architectures for their applications.

## 7.4. Practical Recommendations

The study's conclusions suggest that while implementing image processing systems, practitioners and stakeholders should give equal weight to security and performance optimization. To reduce potential dangers and protect sensitive data, it's essential to implement strong encryption algorithms, regularly check vulnerabilities, and incorporate hardware-level security measures. Through the integration of these pragmatic suggestions into their operational procedures, establishments can augment the robustness and safety of frameworks, cultivating assurance and faith in their application throughout various fields.

## 7.5. Theoretical Contributions

Our study's theoretical foundations have been strengthened by empirical validation of security considerations and performance metrics in deep learning-based image processing. Our results emphasize the interaction between hardware architectures, software frameworks, and security concerns, which advances theoretical understanding.

## 7.6. Limitations and Future Directions

It is crucial to recognize some limitations even if this study has given insightful information on the security implications and performance metrics of image processing operations on CPU and GPU architectures within deep learning frameworks. One of these is the emphasis on particular frameworks and architectures, which might not adequately represent the range of the ever-changing AI ecosystem. Furthermore, the study mainly tackled security issues from a broad viewpoint, providing an opportunity for more in-depth examinations of possible weaknesses. In order to further advance our understanding and implementation of secure systems for image processing tasks, future research could examine the effectiveness of additional frameworks, delve into more detailed security analyses, and address emerging challenges in hardware-based attacks and energy-constrained environments.

# 8. FUTURE DIRECTIONS FOR RESEARCH

## 8.1. Introduction to Future Directions

Future research priorities must prioritize tackling the new opportunities and problems in protecting CPU and GPU architectures inside deep learning frameworks, as the integration of deep learning techniques with image processing activities continues to expand. Researchers may concentrate on strengthening the resilience of AI systems against changing threats, investigating new defense mechanisms, and embracing cutting-edge technologies like blockchain and federated learning by utilizing the insights gained from this study (Suresh Babu, C. V. & Abhinaba Pal. 2023). This proactive strategy will benefit the creation of resilient AI systems with wide-ranging implications for military, security, and healthcare, as well as the field of secure image processing.

## 8.2. Identification of Research Gaps

There are still a number of important research gaps, despite the fact that the current paper offers a thorough investigation of security implications in image processing jobs on CPU and GPU architectures within deep learning frameworks. More research is specifically required to develop new protection strategies against new threats such as adversarial attacks that are specifically designed for deep learning-based image processing systems. To create comprehensive solutions that strike a balance between robustness and efficiency, it is also crucial to look into the trade-offs between security and performance optimization techniques.

## 8.3. Emerging Trends and Technologies

New developments in the field of secure image processing inside deep learning frameworks include blockchain integration, federated learning, and differential privacy (Suresh Babu, C. V. & Sanjai Das 2023). Federated learning is a viable method for secure image processing in decentralized situations by allowing collaborative model training among dispersed devices while maintaining data privacy. By introducing noise to data during model training, differential privacy techniques offer strong privacy guarantees while reducing the possibility of sensitive data leakage. Additionally, the incorporation of blockchain technology enables provenance tracking and secure data sharing, guaranteeing the integrity and traceability of image processing processes among many stakeholders and dissimilar systems. These new developments offer fresh chances to improve the security and privacy of AI-powered image processing systems across a range of uses.

## 8.4. Integration with Existing Literature

For future research directions in safe image processing operations on CPU and GPU architectures inside deep learning frameworks, integration with the body of existing literature is critical. By combining the results of earlier research on hardware vulnerabilities, image processing, and deep learning security, researchers can expand on the body of knowledge and create new directions for advancement. Furthermore, a comprehensive knowledge of the intricate interactions between security, performance, and efficiency in AI-driven image processing systems is made possible by incorporating insights from related domains like cybersecurity, hardware design, and privacy-preserving technologies.

## 8.5. Methodological Considerations

To ensure the validity and relevance of results, methodological considerations are essential for future research in the field of safe image processing operations on CPU and GPU architectures with deep learning frameworks. To improve the generalizability of findings, researchers should place a high priority on using solid experimental designs that cover a variety of datasets and real-world scenarios. To promote reproducibility within the research community and facilitate comparative analysis, standardized benchmarks and evaluation measures ought to be created. Furthermore, given the dynamic nature of the study field and the changing landscape of deep learning frameworks and hardware architectures, researchers should continue to be aware of the viability and usefulness of proposed research approaches.

## 8.6. Interdisciplinary Perspectives

Future research efforts in safe image processing operations on CPU and GPU architectures inside deep learning frameworks will benefit greatly from an interdisciplinary viewpoint. Working together with specialists in human-computer interface, hardware design, and cybersecurity may provide priceless insights into reducing security threats, making the most use of hardware, and improving user experience. The integration of diverse knowledge enables researchers to develop comprehensive solutions that tackle intricate problems at the confluence of hardware design, security, and artificial intelligence. This ensures the resilience and efficacy of image processing systems across a range of domains.

## 8.7. Implications for Practice and Policy

Future discoveries about safe image processing within deep learning frameworks on CPU and GPU architectures will have a significant impact on policy and practice. From a practical standpoint, these insights will direct the creation and implementation of strong AI systems in various industries, guaranteeing the confidentiality and security of critical information (Suresh Babu, C. V., Sharon Manoj et. al 2023). From a policy standpoint, the results of this study will guide the development of rules and guidelines to reduce the hazards connected to AI-driven image processing, encouraging responsibility and trust in the application of these technologies.

## REFERENCES

Babu, C. S., & Das, S. (2023). Impact of blockchain technology on the stock market. In *Revolutionizing Financial Services and Markets Through FinTech and Blockchain* (pp. 44–59). IGI Global.

Babu, C. S., & Pal, A. (2023). Securing Decentralized Finance Through Blockchain Technologies. In Cyber-Physical Systems and Supporting Technologies for Industrial Automation (pp. 351-372). IGI Global.

Bian, S., Liu, C., & Xu, C. (2019). Secure Deep Learning Inference Acceleration on Heterogeneous Platforms. *2019 IEEE International Symposium on Circuits and Systems (ISCAS)*, 1-5. [DOI: ]DOI: 10.1109/ISCAS.2019.8702189

Cui, C., Chen, W., & Ren, J. (2019). Secure Deep Learning Inference in Data Centers. 2019 IEEE 39th International Conference on Distributed Computing Systems (ICDCS), 1375-1384. [DOI: 10.1109/ICDCS.2019.00137]

Fletcher, C., Lu, Q., & Ren, L. (2019). Privacy-preserving Deep Learning on CPU, GPU, and FPGA. *2019 IEEE International Symposium on Hardware Oriented Security and Trust (HOST)*, 35-40. [DOI: ]DOI: 10.1109/HOST45687.2019.8836930

Jain, A., Iyengar, A., & Singh, K. (2019). Safeguarding Deep Learning with CPU, GPU, and RNN: A Review. 2019 IEEE 2nd International Conference on Sustainable Technologies for Industry 4.0 (STI), 1-6. [DOI: 10.1109/STI47621.2019.9113312]

Jin, X., Lu, W., & Li, H. (2019). TrustZone-Assisted Secure Deep Learning on Mobile Devices. *2019 IEEE International Symposium on Hardware Oriented Security and Trust (HOST)*, 158-165. [DOI: ]DOI: 10.1109/HOST45687.2019.8836906

Kasherman, D., Sheng, W., & Devadas, S. (2020). Tornado: Enabling Secure End-to-End Deep Learning via Cryptographically Secure Processors. *2020 IEEE International Symposium on Hardware Oriented Security and Trust (HOST)*, 1-10. [DOI: ]DOI: 10.1109/HOST45687.2020.9126784

Kim, J., Lee, S., & Kim, D. (2018). SGX-LAP: A Secure Deep Learning Accelerator Framework on Intel SGX. *2018 IEEE International Symposium on High-Performance Computer Architecture (HPCA)*, 607-618. [DOI: ]DOI: 10.1109/HPCA.2018.00056

Kim, Y., Lim, D., Yoon, J., & Kim, H. (2019). Secure Deep Learning Inference on Mobile Platforms. 2019 IEEE International Conference on Big Data (Big Data), 5070-5075. [DOI: 10.1109/BigData47090.2019.9006377]

Suresh Babu, C. V., Abirami, S., & Sharon Manoj (2023). AI-Based Carthage Administration Towards Smart City. In C. Chowdhary, B. Swain, & V. Kumar (Eds.), Investigations in Pattern Recognition and Computer Vision for Industry 4.0 (pp. 1-17). IGI Global. https://doi.org/DOI: 10.4018/978-1-6684-8602-3.ch001

Suresh Babu, C. V., Akshayah, N. S., Maclin Vinola, P., & Janapriyan, R. (2023). IoT-Based Smart Accident Detection and Alert System. In Swarnalatha, P., & Prabu, S. (Eds.), *Handbook of Research on Deep Learning Techniques for Cloud-Based Industrial IoT* (pp. 322–337). IGI Global., DOI: 10.4018/978-1-6684-8098-4.ch019

Wei, S., Zhang, Z., & Kim, M. (2018). Secure and Efficient Inference Engine with Deep Learning. *2018 IEEE International Symposium on Multimedia (ISM)*, 455-459. [DOI: ]DOI: 10.1109/ISM.2018.00092

Yi, X., Yu, S., & Bertino, E. (2018). Secure Multi-Party Deep Learning Computation using Secret Sharing and GPUs. *2018 IEEE International Conference on Parallel & Distributed Processing with Applications (ISPA)*, 677-686. [DOI: ]DOI: 10.1109/ISPA.2018.00115

# Compilation of References


Abuzaghleh, O.. (2015). Noninvasive real-time automated skin lesion analysis system for melanoma early detection and prevention. *IEEE Journal of Translational Engineering in Health and Medicine*, *3*, 1–12. https://doi.org/10.1109/JTEHM.2015.2435517

Acharya, D., Jain, R., Panigrahi, S. S., Sahni, R., Jain, S., Deshmukh, S. P., & Bhardwaj, A. (2021). Multi-class Emotion Classification Using EEG Signals. In Garg, D., Wong, K., Sarangapani, J., & Gupta, S. K. (Eds.), *Advanced Computing* (pp. 474–491). Springer Singapore. DOI: 10.1007/978-981-16-0401-0_38

Adek, R. T., & Ula, M. (2020). A Survey on The Accuracy of Machine Learning Techniques for Intrusion and Anomaly Detection on Public Data Sets. *2020 International Conference on Data Science, Artificial Intelligence, and Business Analytics (DATABIA)*. DOI: 10.1109/DATABIA50434.2020.9190436

Aguirre, G. K., & Zarahn, E., & D′ Esposito, M. (1998). A critique of the use of the Kolmogorov-Smirnov (KS) statistic for the analysis of BOLD fMRI data. *Magnetic Resonance in Medicine*, *39*(3), 500–505.

Ahmad, M., Nadeem, M., Islam, M., Ali, S., Agrawal, A., & Ahmad Khan, R. (2023). Selection of Digital Watermarking Techniques for Medical Image Security by Using the Fuzzy Analytical Hierarchy Process. Recent Advances in Computer Science and Communications (Formerly: Recent Patents on Computer Science), 16(7), 74-80.

Ahmad, M., Ai, D., Xie, G., Qadri, S. F., Song, H., Huang, Y., Wang, Y., & Yang, J. (2019). Deep belief network modeling for automatic liver segmentation. *IEEE Access : Practical Innovations, Open Solutions*, *7*, 20585–20595. DOI: 10.1109/ACCESS.2019.2896961

Ahmed, S., Shaikh, A., Alshahrani, H., Alghamdi, A., Alrizq, M., Baber, J., & Bakhtyar, M. (2021, August 9). Transfer Learning Approach for Classification of Histopathology Whole Slide Images. *Sensors (Basel), 21*(16), 5361. DOI: 10.3390/s21165361 PMID: 34450802

Ahn, B., Moon, D., Kim, H. S., Lee, C., Cho, N. H., Choi, H.-K., Kim, D., Lee, J.-Y., Nam, E. J., Won, D., An, H. J., Kwon, S. Y., Shin, S.-J., Jung, H. R., Kwon, D., Park, H., Kim, M., Cha, Y. J., Park, H., & Park, E. (2024). Histopathologic image–based deep learning classifier for predicting platinum-based treatment responses in high-grade serous ovarian cancer. *Nature Communications*, *15*(1), 4253. DOI: 10.1038/s41467-024-48667-6 PMID: 38762636

Aitken, J. F., Pfitzner, J., & Battistutta, D.. (1996). Reliability of computer image analysis of pigmented skin lesions of Australian adolescents. *Journal of Cancer*, *78*(2), 252–257. https://doi.org/10.1002/(SICI)1097-0215(19961009)78:2<252::AID-IJC10>3.0.CO;2-K

Alarcão, S. M., & Fonseca, M. J. (2019). Emotions Recognition Using EEG Signals: A Survey. *IEEE Transactions on Affective Computing*, *10*(3), 374–393. DOI: 10.1109/TAFFC.2017.2714671

Alhussan, A. A., Abdallah, H. A., Alsodairi, S., & Ateya, A. A. (2023). Hybrid Watermarking and Encryption Techniques for Securing Medical Images. *Computer Systems Science and Engineering*, *46*(1), 403–416. DOI: 10.32604/csse.2023.035048

Ali, Z., & Hussain, M. (2022). A novel fragile zero-watermarking algorithm for digital medical images. *Electronics (Basel)*, *11*(5), 710. DOI: 10.3390/electronics11050710

Alquhayz, H., & Raza, B. (2022). Watermarking techniques for the security of medical images and image sequences. *Arabian Journal for Science and Engineering*, *47*(8), 9471–9488. DOI: 10.1007/s13369-021-06254-7

Alzahrani, A. (2022). Enhanced invisibility and robustness of digital image watermarking based on DWT-SVD. *Applied Bionics and Biomechanics*, *2022*, 2022. DOI: 10.1155/2022/5271600 PMID: 35126659

Amaral, T., Silva, L. M., & Alexandre, L. A.. (2014). Transfer learning using rotated image data to improve deep neural network performance. *International Conference on Image Analysis and Recognition (ICIAR: Image Analysis and Recognition)*, 290–300. https://doi.org/10.1007/978-3-319-11758-4_31

American Cancer Society – ovarian cancer key statistics: https://www.cancer.org/content/dam/CRC/PDF/Public/8773.00.pdf

Amine, K., Fares, K., Redouane, K. M., & Salah, E. (2022). Medical image watermarking for telemedicine application security. *Journal of Circuits, Systems, and Computers*, *31*(05), 2250097. DOI: 10.1142/S0218126622500979

Anwar, S. M., Awan, S., Yousaf, S., & Majid, M. (2018, December). Segmentation of liver tumor for computer aided diagnosis. In *2018 IEEE-EMBS Conference on Biomedical Engineering and Sciences (IECBES)* (pp. 366-370). IEEE. DOI: 10.1109/IECBES.2018.8626682

Apostolidis, K. D., & Papakostas, G. A. (2022). Digital watermarking as an adversarial attack on medical image analysis with deep learning. *Journal of Imaging*, *8*(6), 155. DOI: 10.3390/jimaging8060155 PMID: 35735954

Ari, N., & Ustazhanov, M. (2014). Matplotlib in python. 2014 11th International Conference on Electronics, Computer and Computation (ICECCO),. IEEE, 1-6.

Ascher, D., Dubois, P. F., Hinsen, K., Hugunin, J., & Oliphant, T. (2001). Numerical python.

Aydinalp, C.. (2019). In vitro dielectric properties of rat skin tissue for microwave skin cancer detection. *IEEE Access : Practical Innovations, Open Solutions*, *7*, 115208–115216. https://doi.org/10.1109/ACCESS.2019.2935634

Ayed, B. (2015). proposed a method for detecting text data in video frames through big data analytics. *Pattern Recognition Letters*, *96*, 66–75.

Babu, C. S., & Pal, A. (2023). Securing Decentralized Finance Through Blockchain Technologies. In Cyber-Physical Systems and Supporting Technologies for Industrial Automation (pp. 351-372). IGI Global.

Babu, C. S., & Das, S. (2023). Impact of blockchain technology on the stock market. In *Revolutionizing Financial Services and Markets Through FinTech and Blockchain* (pp. 44–59). IGI Global.

Baby, K. (2014). *Big Data: An Ultimate Solution in Health Care.*

Badrinarayanan, V., Handa, A., & Cipolla, R. (2015). A deep convolutional encoder-decoder architecture for robust semantic pixel-wise labelling. *arXiv preprint arXiv:1505.07293.* https://doi.org/10.48550/arXiv.1505.07293

Balasamy, K., & Shamia, D. (2023). Feature extraction-based medical image watermarking using fuzzy-based median filter. *Journal of the Institution of Electronics and Telecommunication Engineers*, *69*(1), 83–91. DOI: 10.1080/03772063.2021.1893231

Balaska, N., Belmeguenai, A., Goutas, A., Ahmida, Z., & Boumerdassi, S. (2022). Securing Medical Data by Combining Encryption and Robust Blind Medical Image Watermarking Based on Zaslavsky Chaotic Map and DCT Coefficients. *SN Computer Science*, *3*(2), 118. DOI: 10.1007/s42979-021-01012-w

Barrett, P., Hunter, J., Miller, J. T., Hsu, J.-C. & Greenfield, P. matplotlib--A Portable Python Plotting Package. Astronomical data analysis software and systems XIV, 2005. 91.

Bashashati, A. (2023). *UBC ovarian cancer subtype classification and outlier detection*. UBCOCEAN.

Bazgir, O., Mohammadi, Z., & Habibi, S. A. H. (2018). Emotion Recognition with Machine Learning Using EEG Signals. *2018 25th National and 3rd International Iranian Conference on Biomedical Engineering (ICBME)*, 1–5. DOI: 10.1109/ICBME.2018.8703559

Begum, M., & Uddin, M. S. (2020). Analysis of digital image watermarking techniques through hybrid methods. *Advances in Multimedia*, *2020*, 1–12. DOI: 10.1155/2020/7912690

Begum, M., & Uddin, M. S. (2020). Digital image watermarking techniques: A review. *Information (Basel)*, *11*(2), 110. DOI: 10.3390/info11020110

Behara, K., Bhero, E., & Agee, J. T. (2024). An improved skin lesion classification using a hybrid approach with active contour snake model and lightweight attention-guided capsule networks. *Diagnostics (Basel)*, *14*(6), 636. https://doi.org/10.3390/diagnostics14060636

Belazi, A., Abd El-Latif, A. A., & Belghith, S. (2016). A novel image encryption scheme based on substitution-permutation network and chaos. *Signal Processing*, *128*, 155–170. DOI: 10.1016/j.sigpro.2016.03.021

Benyoucef, A., & Hamaouche, M. H. (2023). Region-based medical image watermarking approach for secure EPR transmission applied to e-Health. *Arabian Journal for Science and Engineering*, •••, 1–13.

Bernard, J. (2016). *Python data analysis with pandas. Python Recipes Handbook*. Springer. DOI: 10.1007/978-1-4842-0241-8

Bian, S., Liu, C., & Xu, C. (2019). Secure Deep Learning Inference Acceleration on Heterogeneous Platforms. *2019 IEEE International Symposium on Circuits and Systems (ISCAS)*, 1-5. [DOI: ]DOI: 10.1109/ISCAS.2019.8702189

Big Self-Supervised Models are Strong Semi-Supervised Learners, Chen, Ting and Kornblith, Simon and Swersky, Kevin and Norouzi, Mohammad and Hinton, Geoffrey, arXiv preprint arXiv:2006.10029, 2020

Bonakdari, H., Jamshidi, A., Pelletier, J.-P., Abram, F., Tardif, G., & Martel-Pelletier, J. (2021). A warning machine learning algorithm for early knee osteo-arthritis structural progressor patient screening. *Sage (Atlanta, Ga.)*, *13*. DOI: 10.1177/1759720X21993254 PMID: 33747150

Brahim, A., Jennane, R., Riad, R., Janvier, T., Khedher, L., Toumi, H., & Lespessa-illes, E. (2019). A decision support tool for early detection of knee OsteoArthritis using X-ray imaging and machine learning: Data from the OsteoArthritis Initiative. *Computerized Medical Imaging and Graphics*, *73*, 11–18. DOI: 10.1016/j.comp-medimag.2019.01.007 PMID: 30784984

Breen, J., Allen, K., Zucker, K., Hall, G., Orsi, N., & Ravikumar, N. (2023). Efficient subtyping of ovarian cancer histopathology whole slide images using active sampling in multiple instance learning. 10.48550/arXiv.2302.08867.

Brosch, T., Yoo, Y., & Tang, L.. (2015). Deep convolutional encoder networks for multiple sclerosis lesion segmentation. *International Conference on Medical Image Computing and Computer-Assisted Intervention – MICCAI*, Vol. 9351, 3–11. https://doi.org/10.1007/978-3-319-24574-4_1

Brunetti, A., Altini, N., Buongiorno, D., Garolla, E., Corallo, F., Gravina, M., Bevilacqua, V., & Prencipe, B. (2022). A Machine Learning and Radiomics Approach in Lung Cancer for Predicting Histological Subtype. *Applied Sciences (Basel, Switzerland)*, *12*(12), 5829. DOI: 10.3390/app12125829

Buckland, M., & Gey, F. (1994). The relationship between recall and precision. *Journal of the American Society for Information Science*, *45*(1), 12–19. DOI: 10.1002/(SICI)1097-4571(199401)45:1<12::AID-ASI2>3.0.CO;2-L

Budak, Ü., Guo, Y., Tanyildizi, E., & Şengür, A. (2020). Cascaded deep convolutional encoder-decoder neural networks for efficient liver tumor segmentation. *Medical Hypotheses*, *134*, 109431. DOI: 10.1016/j.mehy.2019.109431 PMID: 31669758

Bumrungkun, P.. (2018). Detection skin cancer using SVM and snake model. *2018 International Conference on Digital Arts, Media and Technology (ICDAMT)*, 46–49. https://doi.org/10.1109/ICDAMT.2018.8376521

Burns, C., Bollard, B., & Narayanan, A. (2022). Machine-learning for mapping and monitoring shallow coral reef habitats. *Remote Sensing (Basel)*, *14*(2666), 2666. Advance online publication. DOI: 10.3390/rs14112666

C. K., Prasad, M. L., Kiran, A., Shaker Reddy, P. C., Babu, G. C., & Partheeban, N. (2023). Skin cancer classification using a hybrid convolutional neural network with SVM classifier. *2023 Global Conference on Information Technologies and Communications (GCITC)*, 1–6. https://doi.org/10.1109/GCITC60406.2023.10426054

Cancer Council Australia. (2010). Cancer Council to launch new research/failure to monitor highlights cancer risk. SunSmart/ Skin Cancer Facts and Figures. http://www.cancer.org.au/cancersmartlifestyle/SunSmart/Skincancer-factsandfigures.htm

Chacko, A., & Chacko, S. (2022). Deep learning based robust medical image watermarking exploiting DCT and Harris hawks optimization. *International Journal of Intelligent Systems*, *37*(8), 4810–4844. DOI: 10.1002/int.22742

Chang, Y., Stanley, R. J., & Moss, R. H.. (2005). A systematic heuristic approach for feature selection for melanoma discrimination using clinical images. *Skin Research and Technology : Official Journal of International Society for Bioengineering and the Skin (ISBS) and International Society for Digital Imaging of Skin (ISDIS) and International Society for Skin Imaging (ISSI)*, *11*(3), 165–178. https://doi.org/10.1111/j.1600-0846.2005.00133.x

Chen, L., . . .. (2020). A benchmark dataset for both underwater image enhancement and underwater object detection. arXiv Preprint arXiv:2006.15789. https://doi.org//arXiv.2006.15789DOI: 10.48550

Chen, H., Zhang, Y., Kalra, M. K., Lin, F., Chen, Y., Liao, P., Zhou, J., & Wang, G. (2017). Low-dose CT with a residual encoder-decoder convolutional neural network. *IEEE Transactions on Medical Imaging*, *36*(12), 2524–2535. DOI: 10.1109/TMI.2017.2715284 PMID: 28622671

Chen, X., Zhang, Y., Li, D., & Duan, Q. (2023). Chinese mitten crab detection and gender classification method based on GMNet-YOLOv4. *Computers and Electronics in Agriculture*, *214*, 108318. Advance online publication. DOI: 10.1016/j.compag.2023.108318

Chlebus, G., Meine, H., Moltz, J. H., & Schenk, A. (2017). Neural network-based automatic liver tumor segmentation with random forest-based candidate filtering. *arXiv preprint arXiv:1706.00842*.

Chlebus, G., Schenk, A., Moltz, J. H., van Ginneken, B., Hahn, H. K., & Meine, H. (2018). Automatic liver tumor segmentation in CT with fully convolutional neural networks and object-based postprocessing. *Scientific Reports*, *8*(1), 15497. DOI: 10.1038/s41598-018-33860-7 PMID: 30341319

Christ, P. F., Elshaer, M. E. A., Ettlinger, F., Tatavarty, S., Bickel, M., Bilic, P., & Menze, B. H. (2016). Automatic liver and lesion segmentation in CT using cascaded fully convolutional neural networks and 3D conditional random fields. In *Medical Image Computing and Computer-Assisted Intervention–MICCAI 2016: 19th International Conference, Athens, Greece, October 17-21, 2016* [Springer International Publishing.]. *Proceedings*, *19*(Part II), 415–423.

Craik, A., He, Y., & Contreras-Vidal, J. L. (2019). Deep learning for electroencephalogram (EEG) classification tasks: A review. *Journal of Neural Engineering*, *16*(3), 031001. DOI: 10.1088/1741-2552/ab0ab5 PMID: 30808014

Cueva, J. H. (2022). Detection and Classification of Knee Osteoarthritis. *MDPI*, 2362.

Cui, C., Chen, W., & Ren, J. (2019). Secure Deep Learning Inference in Data Centers. 2019 IEEE 39th International Conference on Distributed Computing Systems (ICDCS), 1375-1384. [DOI: 10.1109/ICDCS.2019.00137]

Cui, A., Li, H., Wang, D., Zhong, J., Chen, Y., & Lu, H. (2020). Global, regional prevalence, incidence and risk factors of knee osteoarthritis in population-based studies. *EClinicalMedicine*, *29-30*, 29–30. DOI: 10.1016/j.eclinm.2020.100587 PMID: 34505846

Cui, Z., Zhong, S., Xu, P., Gong, G., & He, Y. (2013). PANDA: A pipeline toolbox for analyzing brain diffusion images. *Frontiers in Human Neuroscience*, *7*, 42. DOI: 10.3389/fnhum.2013.00042 PMID: 23439846

Culjak, I., Abram, D., Pribanic, T., Dzapo, H., & Cifrek, M. (2012, May). A brief introduction to OpenCV. In 2012 proceedings of the 35th international convention MIPRO (pp. 1725-1730). IEEE.

Dai, J.. (2016). R-FCN: Object detection via region-based fully convolutional networks. *Proceedings of the 2016 Advances in Neural Information Processing Systems (NeurIPS)*, 379-387.

Daily, J., & Lewis, R. R. (2011, September). Using the Global Arrays Toolkit to Reimplement NumPy for Distributed Computation. In SciPy (pp. 29-35).

Dalal, N., & Triggs, B. (2005). Histograms of oriented gradients for human detection. *Proceedings of the 2005 IEEE Computer Society Conference on Computer Vision and Pattern Recognition (CVPR)*, 1, 886-893. DOI: 10.1109/CVPR.2005.177

Devi, H. S., & Mohapatra, H. (2023). A novel robust blind medical image watermarking using GWO optimized DWT-DCT-SVD. *Multimedia Tools and Applications*, *82*(26), 41267–41286. DOI: 10.1007/s11042-023-15158-7

Dey, A., Chowdhuri, P., Pal, P., & Tzu-Chuen, L. (2023, March). SVD-Based Watermarking Scheme for Medical Image Authentication. In International Conference on Network Security and Blockchain Technology (pp. 135-146). Singapore: Springer Nature Singapore.

Dietterich, T. G., Lathrop, R. H., & Lozano-Pérez, T. (1997). Solving the multiple instance problem with axis-parallel rectangles. *Artificial Intelligence*, *89*(1-2), 31–71.

Dillon, J. V., Langmore, I., Tran, D., Brevdo, E., Vasudevan, S., Moore, D., Patton, B., Alemi, A., Hoffman, M., & Saurous, R. A. (2017). Tensorflow distributions. *arXiv preprint arXiv:1711.10604*.

Doersch, C., Gupta, A., & Efros, A. A. (2015). Unsupervised visual representation learning by context prediction. In *Proceedings of the IEEE international conference on computer vision* (pp. 1422-1430).

Dong, C., Loy, C. C., He, K., & Tang, X. (2015). Image super-resolution using deep convolutional networks. *IEEE Transactions on Pattern Analysis and Machine Intelligence*, *38*(2), 295–307. DOI: 10.1109/TPAMI.2015.2439281 PMID: 26761735

Dosselmann, R., & Yang, X. D. (2006). Existing and emerging image quality metrics. *Proceedings of the 2006 Canadian Conference on Electrical and Computer Engineering (CCECE)*, 1906-1913. DOI: 10.1109/CCECE.2006.277467

Do, T. T.. (2015). Accessible melanoma detection using smartphones and mobile image analysis. *IEEE Transactions on Computational Imaging*, *1*(3), 176–184. https://doi.org/10.1109/TCI.2015.2476455

Du, Y. (2018). A Novel Method to Predict Knee Osteoarthritis Progression on MRI Using Machine Learning Methods. *IEEE*, 228-236.

Du, Y. (2018). Knee osteoarthritis severity level classification using whole knee cartilage damage Index and ANN. *CHASE '18: Proceedings, International Conference on Connected Health: Applications, Systems and Engineering Technologies*. IEEE/ACM. DOI: 10.1145/3278576.3278585

Du, Y., Almajalid, R., Shan, J., & Zhang, M. (2018). A Novel Method to Predict Knee Osteoarthritis Progression on MRI Using Machine Learning Methods. *IEEE Transactions on Nanobioscience*, *17*(3), 228–236. DOI: 10.1109/TNB.2018.2840082 PMID: 29994316

Dwivedi, R., Dave, D., Naik, H., Singhal, S., Omer, R., Patel, P., Qian, B., Wen, Z., Shah, T., Morgan, G., & Ranjan, R. (2023). Explainable AI (XAI): Core ideas, techniques, and solutions. *ACM Computing Surveys*, *55*(9), 1–33. DOI: 10.1145/3561048

El-Ghany, S. A. (2023). A fully automatic fine tuned deep learning model for knee osteoarthritis detection and progression analysis. *Egyptian Informatics Journal*, 229-240.

Erat, K., Şahin, E. B., Doğan, F., Merdanoğlu, N., Akcakaya, A., & Durdu, P. O. (2024). Emotion recognition with EEG-based brain-computer interfaces: A systematic literature review. *Multimedia Tools and Applications*, *83*(33), 79647–79694. Advance online publication. DOI: 10.1007/s11042-024-18259-z

Erdt, M., Steger, S., Kirschner, M., & Wesarg, S. (2010, October). Fast automatic liver segmentation combining learned shape priors with observed shape deviation. In *2010 IEEE 23rd International Symposium on Computer-Based Medical Systems (CBMS)* (pp. 249-254). IEEE. DOI: 10.1109/CBMS.2010.6042650

Er, M. J., Chen, J., Zhang, Y., & Gao, W. (2023). Research challenges, recent advances, and popular datasets in deep learning-based underwater marine object detection: A review. *Sensors (Basel)*, *23*(4), 1990. Advance online publication. DOI: 10.3390/s23041990 PMID: 36850584

Esteban, O., Markiewicz, C. J., Blair, R. W., Moodie, C. A., Isik, A. I., Erramuzpe, A., Kent, J. D., Goncalves, M., Dupre, E., Snyder, M., Oya, H., Ghosh, S. S., Wright, J., Durnez, J., Poldrack, R. A., & Gorgolewski, K. J. (2019). fMRIPrep: A robust preprocessing pipeline for functional MRI. *Nature Methods*, *16*(1), 111–116. DOI: 10.1038/s41592-018-0235-4 PMID: 30532080

Evsutin, O., Melman, A., & Meshcheryakov, R. (2020). Digital steganography and watermarking for digital images: A review of current research directions. *IEEE Access : Practical Innovations, Open Solutions*, *8*, 166589–166611. DOI: 10.1109/ACCESS.2020.3022779

Fan, B., Li, Z., & Gao, J. (2022). DwiMark: A multiscale robust deep watermarking framework for diffusion-weighted imaging images. *Multimedia Systems*, *28*(1), 295–310. DOI: 10.1007/s00530-021-00835-0

Fan, Y., Li, J., Bhatti, U. A., Shao, C., Gong, C., Cheng, J., & Chen, Y. (2023). A multi-watermarking algorithm for medical images using inception v3 and dct. *CMC-Computers Materials & Continua*, *74*(1), 1279–1302. DOI: 10.32604/cmc.2023.031445

Farabet, C., Couprie, C., & Najman, L.. (2013). Learning hierarchical features for scene labeling. *IEEE Transactions on Pattern Analysis and Machine Intelligence*, *35*(8), 1915–1929. DOI: 10.1109/TPAMI.2012.231

Farahani, H., Boschman, J., Farnell, D., Darbandsari, A., Zhang, A., Ahmadvand, P., & Bashashati, A. (2022). Deep learning-based histotype diagnosis of ovarian carcinoma whole-slide pathology images. *Modern Pathology*, *35*(12), 1983–1990.

Fassihi, N., Shanbehzadeh, J., & Sarrafzadeh, A.. (2011). Melanoma diagnosis by the use of wavelet analysis based on morphological operators. *Proceedings of the International Multiconference of Engineers and Computer Scientists*, 16–18.

Filiot, A., Ghermi, R., Olivier, A., Jacob, P., Fidon, L., Mac Kain, A., Saillard, C., & Schiratti, J.-B. Scaling Self-Supervised Learning for Histopathology with Masked Image Modeling. *medRxiv* 2023.07.21.23292757; DOI: 10.1101/2023.07.21.23292757

Fletcher, C., Lu, Q., & Ren, L. (2019). Privacy-preserving Deep Learning on CPU, GPU, and FPGA. *2019 IEEE International Symposium on Hardware Oriented Security and Trust (HOST)*, 35-40. [DOI: ]DOI: 10.1109/HOST45687.2019.8836930

Gallego, J. (2021). Cross-scale feature fusion for object detection in optical remote sensing images. *IEEE Geoscience and Remote Sensing Letters*, *18*(3), 431–435. DOI: 10.1109/LGRS.2020.2975541

Garcia, R., Nicosevici, T., & Cufí, X. (2002). On the way to solve lighting problems in underwater imaging. OCEANS'02 MTS/IEEE Conference, 2, 1010-1017. DOI: 10.1109/OCEANS.2002.1192107

Gelenbe, E. (1993). Learning in the Recurrent Random Neural Network. *Neural Computation*, *5*(1), 154–164. DOI: 10.1162/neco.1993.5.1.154

Girshick, R. (2015). Fast R-CNN. *Proceedings of the 2015 IEEE International Conference on Computer Vision (ICCV)*, 1440-1448. DOI: 10.1109/ICCV.2015.169

Gomes, D., Saifuddin Saif, A. F. M., & Nandi, D. (2020). Robust underwater object detection with autonomous underwater vehicle: A comprehensive study. *Proceedings of the 2020 International Conference on Computing Advancements (ICCA)*, 1-8. DOI: 10.1145/3377049.3377052

Gong, C., Liu, J., Gong, M., Li, J., Bhatti, U. A., & Ma, J. (2022). Robust medical zero-watermarking algorithm based on Residual-DenseNet. *IET Biometrics*, *11*(6), 547–556. DOI: 10.1049/bme2.12100

Gong, X. (2021). Pyramidal graph convolutional network for skeleton-based human action recognition. *IEEE Sensors Journal*, *21*(14), 16183–16191. DOI: 10.1109/JSEN.2021.3075722

Gonzalez-Rivero, M., Beijbom, O., Rodriguez-Ramirez, A., Bryant, D. E. P., Ganase, A., Gonzalez-Marrero, Y., Herrera-Reveles, A., Kennedy, E. V., Kim, C. J. S., Lopez-Marcano, S., Markey, K., Neal, B. P., Osborne, K., Reyes-Nivia, C., Sampayo, E. M., Stolberg, K., Taylor, A., Vercelloni, J., Wyatt, M., & Hoegh-Guldberg, O. (2020). Monitoring of coral reefs using artificial intelligence: A feasible and cost-effective approach. *Remote Sensing (Basel)*, *12*(3), 489. DOI: 10.3390/rs12030489

Goryawala, M., Gulec, S., Bhatt, R., McGoron, A. J., & Adjouadi, M. (2014). A low-interaction automatic 3D liver segmentation method using computed tomography for selective internal radiation therapy. *BioMed Research International*, *2014*, 2014. DOI: 10.1155/2014/198015 PMID: 25105118

Green, A., & Martin, N. (1994). Computer image analysis in the diagnosis of melanoma. *Journal of the American Academy of Dermatology*, *31*(6), 958–964. https://doi.org/10.1016/S0190-9622(94)70237-0

Gui, J., Cong, X., Cao, Y., Ren, W., Zhang, J., Zhang, J., Cao, J., & Tao, D. (2023). A comprehensive survey and taxonomy on single image dehazing based on deep learning. *ACM Computing Surveys*, *55*(13s), 1–37. DOI: 10.1145/3576918

Guillen, G. (2019). *Digital Image Processing with Python and OpenCV. Sensor Projects with Raspberry Pi*. Springer. DOI: 10.1007/978-1-4842-5299-4

Gull, S., & Parah, S. A. (2024). Advances in medical image watermarking: A state of the art review. *Multimedia Tools and Applications*, *83*(1), 1407–1447. DOI: 10.1007/s11042-023-15396-9 PMID: 37362709

Guo, Y. (2020). A force levels and gestures integrated multi-task strategy for neural decoding. *Complex & Intelligent Systems*, *6*(3), 469–478. DOI: 10.1007/s40747-020-00140-9

Gupta, M. K., Dadheech, P., Kumar, A., Dogiwal, S. R., Poonia, R. C., Raja, L., & Bhatt, D. P. (2022). Detection and localization for watermarking technique using LSB encryption for DICOM Image. *Journal of Discrete Mathematical Sciences and Cryptography*, *25*(1), 193–204. DOI: 10.1080/09720529.2021.2009193

Guptha, M. N.. (2022). Generative adversarial networks for unmanned aerial vehicle object detection with fusion technology. *Journal of Advanced Transportation*, *2022*, 5798423. Advance online publication. DOI: 10.1155/2022/7111248

Hagedorn, S., Kläbe, S., & Sattler, K.-U. (2021). *Putting Pandas in a Box*. CIDR.

Handley, K. F., Sims, T. T., Bateman, N. W., Glassman, D., Foster, K. I., Lee, S., Yao, J., Yao, H., Fellman, B. M., Liu, J., Lu, Z., Conrads, K. A., Hood, B. L., Barakat, W., Zhao, L., Zhang, J., Westin, S. N., Celestino, J., Rangel, K. M., & Sood, A. K. (2022). Classification of High-Grade Serous Ovarian Cancer Using Tumor Morphologic Characteristics. *JAMA Network Open*, *5*(10), e2236626. DOI: 10.1001/jamanetworkopen.2022.36626 PMID: 36239936

Han, F., Yao, J., Zhu, H., & Wang, C. (2020). Underwater image processing and object detection based on deep CNN method. Hindawi. *Journal of Sensors*, *6740312*, 1–20. Advance online publication. DOI: 10.1155/2020/6707328

Harris, C. R., Millman, K. J., Van Der Walt, S. J., Gommers, R., Virtanen, P., Cournapeau, D., Wieser, E., Taylor, J., Berg, S., Smith, N. J., Kern, R., Picus, M., Hoyer, S., Van Kerkwijk, M. H., Brett, M., Haldane, A., Del Río, J. F., Wiebe, M., Peterson, P., & Oliphant, T. E. (2020). Array programming with NumPy. *Nature*, *585*(7825), 357–362. DOI: 10.1038/s41586-020-2649-2 PMID: 32939066

Hatoum, M. W., Couchot, J. F., Couturier, R., & Darazi, R. (2021). Using deep learning for image watermarking attack. *Signal Processing Image Communication*, *90*, 116019. DOI: 10.1016/j.image.2020.116019

Havaei, M., Davy, A., & Warde-Farley, D.. (2014). Brain tumor segmentation with deep neural networks. *Proceedings of BRATS-MICCAI.*

He, D. M., & Seet, G. G. L. (2002). Underwater lidar imaging in highly turbid waters. *Proceedings of SPIE - the International Society for Optical Engineering*, *4488*, 108–119. DOI: 10.1117/12.452805

Heimann, T., Van Ginneken, B., Styner, M. A., Arzhaeva, Y., Aurich, V., Bauer, C., Beck, A., Becker, C., Beichel, R., Bekes, G., Bello, F., Binnig, G., Bischof, H., Bornik, A., Cashman, P., Ying Chi, , Cordova, A., Dawant, B. M., Fidrich, M., & Wolf, I. (2009). Comparison and evaluation of methods for liver segmentation from CT datasets. *IEEE Transactions on Medical Imaging*, *28*(8), 1251–1265. DOI: 10.1109/TMI.2009.2013851 PMID: 19211338

He, K.. (2017). Mask R-CNN. *Proceedings of the 2017 IEEE International Conference on Computer Vision (ICCV)*, 2961-2969. DOI: 10.1109/ICCV.2017.322

Hochreiter, S., & Schmidhuber, J. (1997). Long Short-Term Memory. *Neural Computation*, *9*(8), 1735–1780. DOI: 10.1162/neco.1997.9.8.1735 PMID: 9377276

Horasan, F., Ali Pala, M., Durdu, A., Akgül, A., Akmeşe, Ö. F., & Yıldız, M. Z. (2022). DWT-SVD based watermarking for high-resolution medical holographic images. *Complexity*, *2022*(1), 2022. DOI: 10.1155/2022/3154650

Horwitz, B. (2003). The elusive concept of brain connectivity. *NeuroImage*, *19*(2), 466–470. DOI: 10.1016/S1053-8119(03)00112-5 PMID: 12814595

Hou, Q., Zhou, D., & Feng, J. (2021). Coordinate attention for efficient mobile network design. In *Proceedings of the IEEE/CVF conference on computer vision and pattern recognition* (pp. 13713-13722).

Houssein, E. H., Abdelkareem, D. A., & Hu, G. (2024). An effective multiclass skin cancer classification approach based on deep convolutional neural network. *Cluster Computing*. Advance online publication. https://doi.org/10.1007/s10586-024-04540-1

Hou, X., Liu, J., Sun, Z., & Li, W. (2019). Underwater image enhancement via Retinex with illumination map transmission. *IEEE Access : Practical Innovations, Open Solutions*, *7*, 145207–145220. DOI: 10.1109/ACCESS.2019.2947384

Huang, J., & Jain, V. (2013). Deep and wide multiscale recursive networks for robust image labeling. *arXiv preprint arXiv:1310.0354*. https://doi.org/10.48550/arXiv.1310.0354

Hua, Z., Zhou, Y., Pun, C. M., & Chen, C. P. (2015). 2D Sine Logistic modulation map for image encryption. *Information Sciences*, *297*, 80–94. DOI: 10.1016/j.ins.2014.11.018

Huntenburg, J., Abraham, A., Loula, J., Liem, F., Dadi, K., & Varoquaux, G. (2017). Loading and plotting of cortical surface representations in Nilearn. *Research Ideas and Outcomes*, *3*, e12342. DOI: 10.3897/rio.3.e12342

Hupé, J.-M. (2015). Statistical inferences under the Null hypothesis: Common mistakes and pitfalls in neuroimaging studies. *Frontiers in Neuroscience*, ●●●, 9. PMID: 25745383

Hu, S., & Liu, T. (2024). Underwater rescue target detection based on acoustic images. *Sensors (Basel)*, *24*(6), 1780. Advance online publication. DOI: 10.3390/s24061780 PMID: 38544042

Ibrahim, A., Gamble, P., Jaroensri, R., Abdelsamea, M. M., Mermel, C. H., Chen, P. H. C., & Rakha, E. A. (2020). Artificial intelligence in digital breast pathology: Techniques and applications. *The Breast : Official Journal of the European Society of Mastology*, *49*, 267–273.

*IR(Infrared) Knee Joint Dataset*. (n.d.). Retrieved 5 28, 2024, from Mrinal Kanti Bhowmik: https://mkbhowmik.in/irD.aspx

Jaber, H. A., Aljobouri, H. K., Çankaya, I., Koçak, O. M., & Algin, O. (2019). Preparing fmri data for postprocessing: Conversion modalities, preprocessing pipeline, and parametric and nonparametric approaches. *IEEE Access : Practical Innovations, Open Solutions*, *7*, 122864–122877. DOI: 10.1109/ACCESS.2019.2937482

Jaffe, J. S. (1990). Computer modeling and the design of optimal underwater imaging systems. *IEEE Journal of Oceanic Engineering*, *15*(2), 101–111. DOI: 10.1109/48.50695

Jain, A., Iyengar, A., & Singh, K. (2019). Safeguarding Deep Learning with CPU, GPU, and RNN: A Review. 2019 IEEE 2nd International Conference on Sustainable Technologies for Industry 4.0 (STI), 1-6. [DOI: 10.1109/STI47621.2019.9113312]

Jamshidi, A., Leclercq, M., Labbe, A., Pelletier, J.-P., Abram, F., Droit, A., & Martel-Pelletier, J. (2020). Identification of the most important features of knee osteoarthritis structural progressors using machine learning methods. *Therapeutic Advances in Musculoskeletal Disease*, *12*, 1759720X20933468. DOI: 10.1177/1759720X20933468 PMID: 32849918

Jerlov, N. G. (1976). *Marine optics*. Elsevier.

Jin, X., Lu, W., & Li, H. (2019). TrustZone-Assisted Secure Deep Learning on Mobile Devices. *2019 IEEE International Symposium on Hardware Oriented Security and Trust (HOST)*, 158-165. [DOI: ]DOI: 10.1109/HOST45687.2019.8836906

Jo, H. J., Gotts, S. J., Reynolds, R. C., Bandettini, P. A., Martin, A., Cox, R. W., & Saad, Z. S. (2013). Effective Preprocessing Procedures Virtually Eliminate Distance-Dependent Motion Artifacts in Resting State FMRI. *Journal of Applied Mathematics*, *2013*, 935154. DOI: 10.1155/2013/935154 PMID: 24415902

*Kaggle*. (n.d.). Retrieved 5 28, 2024, from Knee Osteoarthritis Dataset with Severity Grading: https://www.kaggle.com/datasets/shashwatwork/knee-osteoarthritis-dataset-with-severity

Kang, K., & Wang, X. (2014). Fully convolutional neural networks for crowd segmentation. *arXiv preprint arXiv:1411.4464*. https://doi.org/10.48550/arXiv.1411.4464

Kanimozhi, S., Gayathri, G., & Mala, T. (2019). Multiple real-time object identification using single shot multi-box detection. *Proceedings of the 2019 International Conference on Computational Intelligence in Data Science (ICCIDS)*, 1-6. DOI: 10.1109/ICCIDS.2019.8862041

Kanthavel, R. (2022). *Prediction Model Using Reinforcement Deep Learning Technique for*. Tech Science Press.

Kasherman, D., Sheng, W., & Devadas, S. (2020). Tornado: Enabling Secure End-to-End Deep Learning via Cryptographically Secure Processors. *2020 IEEE International Symposium on Hardware Oriented Security and Trust (HOST)*, 1-10. [DOI: ]DOI: 10.1109/HOST45687.2020.9126784

Kassani, S. H., Kassani, P. H., Wesolowski, M. J., Schneider, K. A., & Deters, R. (2019). Classification of histopathological biopsy images using ensemble of deep learning networks. arXiv preprint arXiv:1909.11870.

Kasture, K. R., Choudhari, D., & Matte, P. N. (2022). Prediction and classification of ovarian cancer using enhanced deep convolutional neural network. *International Journal of Engineering Trends and Technology*, *70*(3), 310–318.

Katsarakis, N. (2021). Detection and recognition of stationary vehicles and seatbelts in intelligent Internet of Things traffic management system. *Neural Computing & Applications*. Advance online publication. DOI: 10.1007/s00521-021-05870-6

Kaur, M., Kaur, J., & Kaur, J. (2011). Survey of contrast enhancement techniques based on histogram equalization. *International Journal of Advanced Computer Science and Applications*, *2*(7), 1–7. DOI: 10.14569/IJACSA.2011.020721

Kelly, R. E.Jr, Alexopoulos, G. S., Wang, Z., Gunning, F. M., Murphy, C. F., Morimoto, S. S., Kanellopoulos, D., Jia, Z., Lim, K. O., & Hoptman, M. J. (2010). Visual inspection of independent components: Defining a procedure for artifact removal from fMRI data. *Journal of Neuroscience Methods*, *189*(2), 233–245. DOI: 10.1016/j.jneumeth.2010.03.028 PMID: 20381530

Keshavarz, A.. (2018). Water-based terahertz metamaterial for skin cancer detection application. *IEEE Transactions on Terahertz Science and Technology*, *8*(4), 397–405. https://doi.org/10.1109/TTHZ.2018.2839273

Kharazmi, P.. (2017). Automated detection and segmentation of vascular structures of skin lesions seen in dermoscopy, with an application to basal cell carcinoma classification. *IEEE Journal of Biomedical and Health Informatics*, *21*(6), 1675–1684. https://doi.org/10.1109/JBHI.2017.2676879

Khashaba, M., Fawzy, M., Abdel-Aziz, A., Eladawei, G., & Nagib, R. (2022). Subtyping of high grade serous ovarian carcinoma: Histopathological and immunohistochemical approach. *Journal of the Egyptian National Cancer Institute*, *34*(1), 6. DOI: 10.1186/s43046-022-00104-9 PMID: 35138498

Kim, Y., Lim, D., Yoon, J., & Kim, H. (2019). Secure Deep Learning Inference on Mobile Platforms. 2019 IEEE International Conference on Big Data (Big Data), 5070-5075. [DOI: 10.1109/BigData47090.2019.9006377]

Kim, A., & Eustice, R. M. (2013). Real-time visual SLAM for autonomous underwater hull inspection using visual saliency. *IEEE Transactions on Robotics*, *29*(3), 719–733. DOI: 10.1109/TRO.2012.2235699

Kim, J., Lee, S., & Kim, D. (2018). SGX-LAP: A Secure Deep Learning Accelerator Framework on Intel SGX. *2018 IEEE International Symposium on High-Performance Computer Architecture (HPCA)*, 607-618. [DOI: ]DOI: 10.1109/HPCA.2018.00056

Korotkov, K.. (2015). A new total body scanning system for automatic change detection in multiple pigmented skin lesions. *IEEE Transactions on Medical Imaging*, *34*(1), 317–325. https://doi.org/10.1109/TMI.2014.2359991

Kumar, A. (2020). A review on implementation of digital image watermarking techniques using LSB and DWT. Information and Communication Technology for Sustainable Development: Proceedings of ICT4SD 2018, 595-602.

Kumar, C. (2023). Hybrid optimization for secure and robust digital image watermarking with DWT, DCT and SPIHT. *Multimedia Tools and Applications*, *83*(11), 1–22. DOI: 10.1007/s11042-023-16903-8

Kumar, C. (2024). Optimized and secure digital image watermarking technique using Henon mapping in redundant domain. *Multimedia Tools and Applications*, *83*(34), 1–16. DOI: 10.1007/s11042-024-18643-9

Kumar, C., Singh, A. K., & Kumar, P. (2018). A recent survey on image watermarking techniques and its application in e-governance. *Multimedia Tools and Applications*, *77*(3), 3597–3622. DOI: 10.1007/s11042-017-5222-8

Kusumaningrum, T. D., Faqih, A., & Kusumoputro, B. (2020). Emotion Recognition Based on DEAP Database using EEG Time-Frequency Features and Machine Learning Methods. *Journal of Physics: Conference Series*, *1501*, 012020. DOI: 10.1088/1742-6596/1501/1/012020

Kuzmin, K. A., & Grekov, N. A. (2021). Developing algorithmic software for automated monitoring of aquatic environment. *Proceedings of the 2021 IEEE Russian Automation Conference (RusAutoCon)*, 1-5. DOI: 10.1109/RusAutoCon52004.2021.9537506

Kwon, S. B. (2020). *Machine Learning-Based Automatic Classification of Knee Osteoarthritis Severity Using Gait Data and Radiographic Images*. IEEE. DOI: 10.1109/ACCESS.2020.3006335

Lakshminarayanan, V., Ravikumar, A., Sriraman, H., Alla, S., & Chattu, V. K. (2023). Health Care Equity Through Intelligent Edge Computing and Augmented Reality/ Virtual Reality: A Systematic Review. *Journal of Multidisciplinary Healthcare*, *16*, 2839–2859. DOI: 10.2147/JMDH.S419923 PMID: 37753339

Lamecker, H., Lange, T., & Seebass, M. (2004). Segmentation of the liver using a 3D statistical shape model.

Latif, G., Iskandar, D. A., Alghazo, J., Butt, M., & Khan, A. H. (2018). Deep CNN based MR image denoising for tumor segmentation using watershed transform. International Journal of Engineering & Technology, 7(2.3), 37-42.

Lawrence, R. M., Bridgeford, E. W., Myers, P. E., Arvapalli, G. C., Ramachandran, S. C., Pisner, D. A., Frank, P. F., Lemmer, A. D., Nikolaidis, A., & Vogelstein, J. T. (2021). Standardizing human brain parcellations. *Scientific Data*, *8*(1), 78. DOI: 10.1038/s41597-021-00849-3 PMID: 33686079

Lazard, T., Bataillon, G., Naylor, P., Popova, T., Bidard, F. C., Stoppa-Lyonnet, D., Stern, M. H., Decencière, E., Walter, T., & Vincent-Salomon, A. (2022, December 20). Deep learning identifies morphological patterns of homologous recombination deficiency in luminal breast cancers from whole slide images. *Cell Reports Medicine*, *3*(12), 100872. DOI: 10.1016/j.xcrm.2022.100872 PMID: 36516847

LeCun, B., Bottou, L., & Bengio, Y.. (1998). Gradient-based learning applied to document recognition. *Proceedings of the IEEE*, *86*(11), 2278–2324. https://doi .org/10.1109/5.726791

LeCun, Y., Bengio, Y., & Hinton, G. (2015). Deep learning. *Nature*, *521*(7553), 436–444. DOI: 10.1038/nature14539 PMID: 26017442

Lee, H. C. (1994). *Skin cancer diagnosis using hierarchical neural networks and fuzzy logic*. Department of Computer Science, University of Missouri.

Lee, J. E., Seo, Y. H., & Kim, D. W. (2020). Convolutional neural network-based digital image watermarking adaptive to the resolution of image and watermark. *Applied Sciences (Basel, Switzerland)*, *10*(19), 6854. DOI: 10.3390/app10196854

Lee, J., Kim, N., Lee, H., Seo, J. B., Won, H. J., Shin, Y. M., & Shin, Y. G. (2007). Efficient liver segmentation exploiting level-set speed images with 2.5 D shape propagation. In *Proceedings of the MICCAI Workshop on 3-D Segmentat. Clinic: A Grand Challenge* (pp. 189-196).

Lemenkova, P. (2019). Processing oceanographic data by Python libraries NumPy, SciPy and Pandas. *Aquatic Research*, *2*, 73–91. DOI: 10.3153/AR19009

Liang, J. (2018). Recurrent attention network using spatial- temporal relations for action recognition. *Signal Processing*, *145*, 137–145. DOI: 10.1016/j.sig-pro.2017.12.008

Li, C., Anwar, S., Hou, J., Cong, R., Guo, C., & Han, J. (2020). An underwater image enhancement benchmark dataset and beyond. *IEEE Transactions on Image Processing*, *29*, 4376–4389. DOI: 10.1109/TIP.2019.2955241 PMID: 31796402

Li, D., Chen, Y. W., Li, J., Cao, L., Bhatti, U. A., & Zhang, P. (2022). Robust watermarking algorithm for medical images based on accelerated-KAZE discrete cosine transform. *IET Biometrics*, *11*(6), 534–546. DOI: 10.1049/bme2.12102

Li, G., Chen, X., Shi, F., Zhu, W., Tian, J., & Xiang, D. (2015). Automatic liver segmentation based on shape constraints and deformable graph cut in CT images. *IEEE Transactions on Image Processing*, *24*(12), 5315–5329. DOI: 10.1109/TIP.2015.2481326 PMID: 26415173

Linguraru, M. G., Richbourg, W. J., Watt, J. M., Pamulapati, V., & Summers, R. M. (2012). Liver and tumor segmentation and analysis from CT of diseased patients via a generic affine invariant shape parameterization and graph cuts. In *Abdominal Imaging. Computational and Clinical Applications: Third International Workshop, Held in Conjunction with MICCAI 2011, Toronto, ON, Canada, September 18, 2011, Revised Selected Papers 3* (pp. 198-206). Springer Berlin Heidelberg. DOI: 10.1007/978-3-642-28557-8_25

Lin, T. Y., Maire, M., Belongie, S., Hays, J., Perona, P., Ramanan, D., Dollár, P., & Zitnick, C. L. (2014). Microsoft COCO: Common objects in context. *Proceedings of the 2014 European Conference on Computer Vision (ECCV)*, 740-755. DOI: 10.1007/978-3-319-10602-1_48

Liu, J., Meng, H., Nandi, A., & Li, M. (2016). Emotion detection from EEG recordings. *2016 12th International Conference on Natural Computation, Fuzzy Systems and Knowledge Discovery (ICNC-FSKD)*, 1722–1727. DOI: 10.1109/FSKD.2016.7603437

Liu, W., Li, J., Shao, C., Ma, J., Huang, M., & Bhatti, U. A. (2022, July). Robust zero watermarking algorithm for medical images using local binary pattern and discrete cosine transform. In *International Conference on Artificial Intelligence and Security* (pp. 350-362). Cham: Springer International Publishing. DOI: 10.1007/978-3-031-06764-8_28

Liu, Z., Zhuang, Y., Jia, P., Wu, C., Xu, H., & Liu, Z. (2022). A novel underwater image enhancement algorithm and an improved underwater biological detection pipeline. *Journal of Marine Science and Engineering*, *10*(9), 1204. Advance online publication. DOI: 10.3390/jmse10091204

Li, Y., Li, J., Shao, C., Bhatti, U. A., & Ma, J. (2022, July). Robust multi-watermarking algorithm for medical images using patchwork-DCT. In *International Conference on Artificial Intelligence and Security* (pp. 386-399). Cham: Springer International Publishing. DOI: 10.1007/978-3-031-06791-4_31

Li, Y., Liu, J., Li, X., & Tang, C. (2023). Underwater object detection algorithm based on adding channel and spatial fusion attention mechanism. *MDPI, 11*(6), 822. DOI: 10.3390/mdpi11060822

Long, E. S., Shelhamer, E., & Darrell, T. (2015). Fully convolutional networks for semantic segmentation. *Proceedings of the IEEE Conference on Computer Vision and Pattern Recognition*, 3431–3440. https://doi.org/10.1109/CVPR.2015.7298965

Long, J., Shelhamer, E., & Darrell, T. (2015). Fully convolutional networks for semantic segmentation. *IEEE Conference on Computer Vision and Pattern Recognition*, 3431–3440. https://doi.org/10.1109/CVPR.2015.7298965

Lopez-Vazquez, V., Lopez-Guede, J. M., Chatzievangelou, D., & Aguzzi, J. (2023). Deep learning based deep-sea automatic image enhancement and animal species classification. *Journal of Big Data, 10*(1), 37. Advance online publication. DOI: 10.1186/s40537-023-00711-w

Lu, H., & Serikawa, S. (2014). Underwater scene enhancement using weighted guided median filter. *Proceedings of the 2014 IEEE International Conference on Multimedia and Expo (ICME)*, 1-6. DOI: 10.1109/ICME.2014.6890157

Ma, B., Guo, Y., Hu, W., Yuan, F., Zhu, Z., Yu, Y., & Zou, H. (2020, October 2). Artificial Intelligence-Based Multiclass Classification of Benign or Malignant Mucosal Lesions of the Stomach. *Frontiers in Pharmacology, 11*, 572372. DOI: 10.3389/fphar.2020.572372 PMID: 33132910

Magalhães, R., Marques, P., Veloso, T., Soares, J. M., Sousa, N., & Alves, V. (2015). Construction of functional brain connectivity networks. In Distributed Computing and Artificial Intelligence, 12th International Conference (pp. 303-310). Springer International Publishing.

Magdy, M., Ghali, N. I., Ghoniemy, S., & Hosny, K. M. (2022). Multiple zero-watermarking of medical images for internet of medical things. *IEEE Access : Practical Innovations, Open Solutions, 10*, 38821–38831. DOI: 10.1109/ACCESS.2022.3165813

Mahum, R. (2021). A Novel Hybrid Approach Based on Deep CNN Features to Detect Knee Osteoarthritis. *MDPI*, 6189.

Mäkynen, M., & Karvonen, J. (2017). Incidence angle dependence of first-year sea ice backscattering coefficient in Sentinel-1 SAR imagery over the Kara Sea. *IEEE Transactions on Geoscience and Remote Sensing*, *55*(11), 6355–6364. DOI: 10.1109/TGRS.2017.2721981

Medwin, H., & Clay, C. S. (1998). *Fundamentals of acoustical oceanography*. Academic Press.

Michielli, N., Caputo, A., Scotto, M., Mogetta, A., Pennisi, O. A. M., Molinari, F., & Salvi, M. (2022). Stain normalization in digital pathology: Clinical multi-center evaluation of image quality. *Journal of Pathology Informatics*, *13*, 100145.

Mirbeik, A.. (2018). Synthetic ultra-high-resolution millimeter-wave imaging for skin cancer detection. *IEEE Transactions on Biomedical Circuits and Systems*, *12*(5), 1051–1061. https://doi.org/10.1109/TBCAS.2018.2858412

Mirbeik, A.. (2018). Ultrawideband, stable normal and cancer skin tissue phantoms for millimeter-wave skin cancer imaging. *IEEE Access : Practical Innovations, Open Solutions*, *10*, 1109. https://doi.org/10.1109/ACCESS.2018.2805312

Moad, M. S., Kafi, M. R., & Khaldi, A. (2022). A wavelet based medical image watermarking scheme for secure transmission in telemedicine applications. *Microprocessors and Microsystems*, *90*, 104490. DOI: 10.1016/j.micpro.2022.104490

Moad, M. S., Kafi, M. R., & Khaldi, A. (2022). Medical image watermarking for secure e-healthcare applications. *Multimedia Tools and Applications*, *81*(30), 44087–44107. DOI: 10.1007/s11042-022-12004-0

Mobley, C. D. (1994). *Light and water: Radiative transfer in natural waters*. Academic Press.

Mobley, C., Stramski, D., Bissett, W., & Boss, E. (2004). Optical modeling of ocean waters: Is the Case 1–Case 2 classification still useful? *Oceanography (Washington, D.C.)*, *17*(2), 60–67. DOI: 10.5670/oceanog.2004.48

Mohanarathinam, A., Kamalraj, S., Prasanna Venkatesan, G. K. D., Ravi, R. V., & Manikandababu, C. S. (2020). Digital watermarking techniques for image security: A review. *Journal of Ambient Intelligence and Humanized Computing*, *11*(8), 3221–3229. DOI: 10.1007/s12652-019-01500-1

Morra, G. & Morra, G. (2018). Fast Python: NumPy and Cython. *Pythonic Geodynamics: Implementations for Fast Computing*, 35-60.

Moustakidis, S., Christodoulou, E., Papageorgiou, E., Kokkotis, C., Papandrianos, N., & Tsaopoulos, D. (2019). Application of machine intelligence for osteoarthritis classification: A classical implementation and a quantum perspective. *Quantum Machine Intelligence*, *1*(3-4), 73–86. DOI: 10.1007/s42484-019-00008-3

Najva and Bijoy. (2016). proposed a model for the detection and classification of objects in videos by combining Tensor features with the SIFT approach. *Infrared Physics & Technology*, *93*, 205–212.

*National Institute of Health*. (n.d.). Retrieved 12 31, 2023, from The Osteoarthritis Initiative: https://nda.nih.gov/oai

Nawaz, S. A., Li, J., Bhatti, U. A., Shoukat, M. U., Li, D., & Raza, M. A. (2024). Hybrid watermarking algorithm for medical images based on digital transformation and MobileNetV2. *Information Sciences*, *653*, 119810. DOI: 10.1016/j.ins.2023.119810

Nawaz, S. A., Li, J., Shoukat, M. U., Bhatti, U. A., & Raza, M. A. (2023). Hybrid medical image zero watermarking via discrete wavelet transform-ResNet101 and discrete cosine transform. *Computers & Electrical Engineering*, *112*, 108985. DOI: 10.1016/j.compeleceng.2023.108985

Ngo, D., Lee, G. D., & Kang, B. (2019). Improved color attenuation prior for single-image haze removal. *Applied Sciences (Basel, Switzerland)*, *9*(19), 4011. Advance online publication. DOI: 10.3390/app9194011

Oliphant, T. E. (2006). *Guide to numpy*. Trelgol Publishing USA.

Olsson, S. (2021). Automating classifcation of osteoarthritis. *BMC Musculoskeletal Disorders*.

Ondrésik, M. (2016). Management of knee osteoarthritis. Current status and future trends. *Biotechnology and Bioengineering*, 717–739. PMID: 27618194

Pallaw, V. K., Singh, K. U., Kumar, A., Singh, T., Swarup, C., & Goswami, A. (2023). A robust medical image watermarking scheme based on nature-inspired optimization for telemedicine applications. *Electronics (Basel)*, *12*(2), 334. DOI: 10.3390/electronics12020334

Pan, S., & Dawant, B. M. (2001, July). Automatic 3D segmentation of the liver from abdominal CT images: a level-set approach. In *Medical Imaging 2001: Image Processing* (Vol. 4322, pp. 128-138). SPIE.

Parker, D. B., & Razlighi, Q. R. (2019). The Benefit of Slice Timing Correction in Common fMRI Preprocessing Pipelines. *Frontiers in Neuroscience*, *13*, 13. DOI: 10.3389/fnins.2019.00821 PMID: 31551667

Parthasarathy, S., & Sankaran, P. (2012). An automated multi scale retinex with color restoration for image enhancement. *Proceedings of the 2012 National Conference on Communications (NCC)*, 1-5. DOI: 10.1109/NCC.2012.6176791

Patel, Z., Desai, C., Tabib, R. A., Bhat, M., Patil, U., & Mudengudi, U. (2020). Framework for underwater image enhancement. *Procedia Computer Science*, *171*, 491–497. DOI: 10.1016/j.procs.2020.04.052

Pedregosa, F., Varoquaux, G., Gramfort, A., Michel, V., Thirion, B., Grisel, O., Blondel, M., Prettenhofer, P., Weiss, R. & Dubourg, V. (2011). Scikit-learn: Machine learning in Python. *the Journal of machine Learning research,* 12, 2825-2830.

Peng, Y. T., Cao, K., & Cosman, P. C. (2018). Generalization of the dark channel prior for single image restoration. *IEEE Transactions on Image Processing*, *27*(6), 2856–2868. DOI: 10.1109/TIP.2018.2813092 PMID: 29570087

Piccolo, C. L., Mallio, C. A., Vaccarino, F., Grasso, R. F., & Zobel, B. B. (2023). Imaging of knee osteoarthritis: A review of multimodal diagnostic approach. *Quantitative Imaging in Medicine and Surgery*, *13*(11), 7582–7595. DOI: 10.21037/qims-22-1392 PMID: 37969633

Pinheiro, P., & Collobert, R. (2014). Recurrent convolutional neural networks for scene labeling. *Proceedings of the 31st International Conference on Machine Learning*, 82–90.

Pizer, S. M., Amburn, E. P., Austin, J. D., Cromartie, R., Geselowitz, A., Greer, T., ter Haar Romeny, B., Zimmerman, J. B., & Zuiderveld, K. (1987). Adaptive histogram equalization and its variations. *Computer Vision Graphics and Image Processing*, *39*(3), 355–368. DOI: 10.1016/S0734-189X(87)80186-X

Pollard, T. C. (2008). The assessment of early osteoarthritis. *The Bone & Joint*, 411-421.

Pongsakonpruttikul, N., Angthong, C., Kittichai, V., Chuwongin, S., Puengpipattrakul, P., Thongpat, P., & Tongloy, T. (2022). Artificial intelligence assistance in radiographic detection and classification of knee osteoarthritis and its severity: A cross-sectional diagnostic study. *European Review for Medical and Pharmacological Sciences*, *26*(5), 1549–1558.

Prakash, K. B., Ruwali, A., & Kanagachidambaresan, G. (2021). *Introduction to Tensorflow Package. Programming with TensorFlow*. Springer. DOI: 10.1007/978-3-030-57077-4

Priyadharsini, R., Sree Sharmila, T., & Rajendran, V. (2018). A wavelet transform based contrast enhancement method for underwater acoustic images. *Multidimensional Systems and Signal Processing*, *29*(5), 1845–1859. DOI: 10.1007/s11045-017-0533-5

Qu, S., Zhang, H., Wu, W., Xu, W., & Li, Y. (2021). Symmetric pyramid attention convolutional neural network for moving object detection. *Signal, Image and Video Processing*, *15*(8), 1747–1755. DOI: 10.1007/s11760-021-01920-7

Rahmati, M., Mobasheri, A., & Mozafari, M. (2016). Inflammatory mediators in osteoarthritis: A critical review of the state-of-the-art, current prospects, and future challenges. *Bone*, *85*, 81–90. DOI: 10.1016/j.bone.2016.01.019 PMID: 26812612

Rampasek, L., & Goldenberg, A. (2016). TensorFlow: Biology's Gateway to Deep Learning? *Cell Systems*, *2*(1), 12–14. DOI: 10.1016/j.cels.2016.01.009 PMID: 27136685

Ramya and Rajeswari. (2016). the current image and background image to categorize pixels as foreground or background. *IEEE Trans Intell Transp*, *21*(10), 4435–4443.

Ravikumar, A., & Sriraman, H. (2023a). *Acceleration of Image Processing and Computer Vision Algorithms* [Chapter]. Handbook of Research on Computer Vision and Image Processing in the Deep Learning Era; IGI Global. DOI: 10.4018/978-1-7998-8892-5.ch001

Ravikumar, A., & Sriraman, H. (2023b). Mitigating Vanishing Gradient in SGD Optimization in Neural Networks. In Ranganathan, G., Papakostas, G. A., & Rocha, Á. (Eds.), *Inventive Communication and Computational Technologies* (pp. 1–11). Springer Nature Singapore. DOI: 10.1007/978-981-99-5166-6_1

Ravikumar, A., & Sriraman, H. (2024a). A Deep Understanding of Long Short-Term Memory for Solving Vanishing Error Problem: LSTM-VGP. In Baby Maruthi, P., Prasad, S., & Tyagi, A. K. (Eds.), *Machine Learning Algorithms Using Scikit and TensorFlow Environments* (pp. 74–90). IGI Global., DOI: 10.4018/978-1-6684-8531-6.ch004

Ravikumar, A., Sriraman, H., Sai Saketh, P. M., Lokesh, S., & Karanam, A. (2022). Effect of neural network structure in accelerating performance and accuracy of a convolutional neural network with GPU/TPU for image analytics. *PeerJ. Computer Science*, *8*, e909. DOI: 10.7717/peerj-cs.909 PMID: 35494877

Ren, Z., Zhang, Q., Gao, X., Hao, P., & Cheng, J. (2021). Multi-modality learning for human action recognition. *Multimedia Tools and Applications*, *80*(11), 16185–16203. DOI: 10.1007/s11042-019-08576-z

Rezayatmand, M., & Naghsh, A. (2022). Robust and semi-blind digital image watermarking method based on DWT and SVD. *Journal of Intelligent Procedures in Electrical Technology, 13*(51), 1–18.

Riddle, W. R., & Pickens, D. R. (2005). Extracting data from a DICOM file. *Medical Physics, 32*(6Part1), 1537–1541. DOI: 10.1118/1.1916183 PMID: 16013711

Risha and Kumar. (2016). proposed an optic flow with morphological operations for object detection. *IEEE Transactions on Image Processing, 20*(6), 1709–1724.

Ronneberger, O., Fischer, P., & Brox, T. (2015). U-net: Convolutional networks for biomedical image segmentation. *Proceedings of the 18th International Conference on Medical Image Computing and Computer-Assisted Intervention (MICCAI)*, 8. https://doi.org/10.1007/978-3-319-24574-4_28

Rusko, L., Bekes, G., Nemeth, G., & Fidrich, M. (2007). Fully automatic liver segmentation for contrast-enhanced CT images. *MICCAI Wshp. 3D Segmentation in the Clinic: A Grand Challenge, 2*(7), 143-150.

Said, A., & Pearlman, W. A. (1996). A new, fast, and efficient image codec based on set partitioning in hierarchical trees. *IEEE Transactions on Circuits and Systems for Video Technology, 6*(3), 243–250. DOI: 10.1109/76.499834

Salim, M. Z., Abboud, A. J., & Yildirim, R. (2022). A visual cryptography-based watermarking approach for the detection and localization of image forgery. *Electronics (Basel), 11*(1), 136. DOI: 10.3390/electronics11010136

Sanivarapu, P. V. (2022). Adaptive tamper detection watermarking scheme for medical images in transform domain. *Multimedia Tools and Applications, 81*(8), 11605–11619. DOI: 10.1007/s11042-022-12273-9

*Sethu Institute of Technology*. (n.d.). Retrieved 5 28, 2024, from OSTHERSIT: https://sethu.ac.in/osthersit/

Shamrat, J. M., . . .. (2022). Human face recognition applying Haar cascade classifier. In M. Tiwari & S. K. Gupta (Eds.), Pervasive Computing and Social Networking: Proceedings of ICPCSN 2021 (pp. 401-411). Springer. https://doi.org/DOI: 10.1007/978-981-16-4440-1_39

Shao, H.-M., Wang, J.-G., Wang, Y., Yao, Y., & Liu, J. (2019). EEG-Based Emotion Recognition with Deep Convolution Neural Network. *2019 IEEE 8th Data Driven Control and Learning Systems Conference (DDCLS)*, 1225–1229. DOI: 10.1109/DDCLS.2019.8908880

Sheng, M., Li, J., Bhatti, U. A., Liu, J., Huang, M., & Chen, Y. W. (2023). Zero watermarking algorithm for medical image based on Resnet50-DCT. *Computers, Materials & Continua*, *75*(1), 293–309. DOI: 10.32604/cmc.2023.036438

She, Z., Liu, Y., & Damatoa, A. (2007). Combination of features from skin pattern and ABCD analysis for lesion classification. *Skin Research and Technology : Official Journal of International Society for Bioengineering and the Skin (ISBS) and International Society for Digital Imaging of Skin (ISDIS) and International Society for Skin Imaging (ISSI)*, *13*(1), 25–33. https://doi.org/10.1111/j.1600-0846.2007.00177.x

Shivani, J. D., & Senapati, R. K. (2017). Robust image embedded watermarking using DCT and listless SPIHT. *Future Internet*, *9*(3), 33. DOI: 10.3390/fi9030033

Singaravelan, S., Selvakumar, V., Balaganesh, S., Gopalsamy, P., & Arun, R. (2024). Multiple feature fusion in particle filter framework for visual tracking. In *Improving Security, Privacy, and Connectivity Among Telemedicine Platforms* (pp. 238–258). IGI Global. https://doi.org/10.4018/978-1-6684-5958-1.ch010

Singh, K. N., Singh, O. P., Singh, A. K., & Agrawal, A. K. (2022). Singh, K. N., Singh, O. P., Singh, A. K., & Agrawal, A. K. (2022). Watmif: Multimodal medical image fusion-based watermarking for telehealth applications. *Cognitive Computation*, •••, 1–17. PMID: 35818513

Singh, M., & Singh, A. K. (2023). A comprehensive survey on encryption techniques for digital images. *Multimedia Tools and Applications*, *82*(8), 11155–11187. DOI: 10.1007/s11042-022-12791-6

Sinha Roy, S., Basu, A., & Chattopadhyay, A. (2020). On the implementation of a copyright protection scheme using digital image watermarking. *Multimedia Tools and Applications*, *79*(19), 13125–13138. DOI: 10.1007/s11042-020-08652-9

Sinhal, R., & Ansari, I. A. (2020). A multipurpose image watermarking scheme for digital image protection. *International Journal of System Assurance Engineering and Management*, *11*(2), 274–286. DOI: 10.1007/s13198-019-00855-0

Sinhal, R., & Ansari, I. A. (2023). Machine learning based multipurpose medical image watermarking. *Neural Computing & Applications*, *35*(31), 23041–23062. DOI: 10.1007/s00521-023-08457-5 PMID: 37362569

Sinhal, R., Sharma, S., Ansari, I. A., & Bajaj, V. (2022). Multipurpose medical image watermarking for effective security solutions. *Multimedia Tools and Applications*, *81*(10), 14045–14063. DOI: 10.1007/s11042-022-12082-0 PMID: 35233177

Sivaraman, R., Padmaa, M., Sridevi, A., Raj, V., Manikandan, V., Siva, J., Bhogi, T., & Rengarajan, A. (2023). Pullikolam assisted medical image watermarking on reconfigurable hardware. *Multimedia Tools and Applications*, *82*(14), 21193–21203. DOI: 10.1007/s11042-023-14725-2

Song, H., Zhang, Q., & Wang, S. (2014, November). Liver segmentation based on SKFCM and improved GrowCut for CT images. In *2014 IEEE international conference on bioinformatics and biomedicine (BIBM)* (pp. 331-334). IEEE. DOI: 10.1109/BIBM.2014.6999179

Soualmi, A., Alti, A., & Laouamer, L. (2022). An imperceptible watermarking scheme for medical image tamper detection. [IJISP]. *International Journal of Information Security and Privacy*, *16*(1), 1–18. DOI: 10.4018/IJISP.2022010102

Soundrapandiyan and Mouli. (2015) proposed a novel and adaptive method for pedestrian detection: *Proceedings of the IEEE winter conference on applications of computer vision (WACV)*, pp 2774–2783

Spitaels, D. (2020). Epidemiology of knee osteoarthritis in general practice: A registry-based study. *BMJ (Clinical Research Ed.)*. PMID: 31964664

Sung, M., Yu, S. C., & Girdhar, Y. (2017). Vision based real-time fish detection using convolutional neural network. *Proceedings of the 2017 OCEANS Conference (Aberdeen)*, 1-5. DOI: 10.1109/OCEANSE.2017.8084889

Suresh Babu, C. V., Abirami, S., & Sharon Manoj (2023). AI-Based Carthage Administration Towards Smart City. In C. Chowdhary, B. Swain, & V. Kumar (Eds.), Investigations in Pattern Recognition and Computer Vision for Industry 4.0 (pp. 1-17). IGI Global. https://doi.org/DOI: 10.4018/978-1-6684-8602-3.ch001

Suresh Babu, C. V., Akshayah, N. S., Maclin Vinola, P., & Janapriyan, R. (2023). IoT-Based Smart Accident Detection and Alert System. In Swarnalatha, P., & Prabu, S. (Eds.), *Handbook of Research on Deep Learning Techniques for Cloud-Based Industrial IoT* (pp. 322–337). IGI Global., DOI: 10.4018/978-1-6684-8098-4.ch019

Tan, Mingxing & Le, Quoc. (2019). EfficientNet: Rethinking Model Scaling for Convolutional Neural Networks.

Tarekegn, A., Alaya Cheikh, F., Ullah, M., Sollesnes, E., Alexandru, C., Azar, S., Erol, E., & Suciu, G. (2023). Underwater object detection using image enhancement and deep learning models. *Proceedings of the 2023 European Workshop on Visual Information Processing (EUVIP)*, 1-6. DOI: 10.1109/EUVIP58404.2023.10323047

Tariq, T. (2023). Knee Osteoarthritis Detection and Classification Using X-Rays. *IEEE*, 48292 - 48303.

Tiwari, A., & Srivastava, V. K. (2024). Image watermarking techniques based on Schur decomposition and various image invariant moments: A review. *Multimedia Tools and Applications, 83*(6), 16447–16483. DOI: 10.1007/s11042-023-16109-y

Tritech International Limited. (2008). *Side scan sonar*. Tritech International Limited.

Tsai, T.-H. (2021). DFFNet: An IoT-perceptive dual feature fusion network for general real-time semantic segmentation. *Information Sciences, 565*, 326–343. DOI: 10.1016/j.ins.2021.02.004

Uludag, K., Wang, D. M., Mohamoud, Y., Wu, H. E., & Zhang, X. (2023). Antipsychotic-based machine learning models may help prediction of tardive dyskinesia in patients with schizophrenia. *Schizophrenia Research, 252*, 33–35. DOI: 10.1016/j.schres.2022.12.026 PMID: 36621324

Urban, B., Bendszus, M., & Hamprecht, F. A.. (2014). Multi-modal brain tumor segmentation using deep convolutional neural networks. *Proceedings of BRATS-MICCAI.*

Vaidya, S. P. (2023). Fingerprint-based robust medical image watermarking in hybrid transform. *The Visual Computer, 39*(6), 2245–2260. DOI: 10.1007/s00371-022-02406-4 PMID: 35125576

Van Eycke, Y. R., Foucart, A., & Decaestecker, C. (2019). Strategies to reduce the expert supervision required for deep learning-based segmentation of histopathological images. *Frontiers in medicine, 6*, 222.

Vancea, F., Mitrea, D., Nedevschi, S., Rotaru, M., Stefanescu, H., & Badea, R. (2019, September). Hepatocellular carcinoma segmentation within ultrasound images using convolutional neural networks. In *2019 IEEE 15th International Conference on Intelligent Computer Communication and Processing (ICCP)* (pp. 483-490). IEEE. DOI: 10.1109/ICCP48234.2019.8959687

Virtanen, P., Gommers, R., Oliphant, T. E., Haberland, M., Reddy, T., Cournapeau, D., Burovski, E., Peterson, P., Weckesser, W., Bright, J., van der Walt, S. J., Brett, M., Wilson, J., Millman, K. J., Mayorov, N., Nelson, A. R. J., Jones, E., Kern, R., Larson, E., & Vázquez-Baeza, Y. (2020). SciPy 1.0: Fundamental algorithms for scientific computing in Python. *Nature Methods, 17*(3), 261–272. DOI: 10.1038/s41592-019-0686-2 PMID: 32015543

Wadhera, S., Kamra, D., Rajpal, A., Jain, A., & Jain, V. (2022). A comprehensive review on digital image watermarking. arXiv preprint arXiv:2207.06909.

Wang, Z., Zhu, J., Yan, T., & Yang, L. (2019). A new modified wavelet-based ECG denoising. Computer Assisted Surgery, 24(sup1), 174-183. DOI: 10.1080/24699322.2018.1560088

Wang, C. W., Chang, C. C., Khalil, M. A., Lin, Y.-J., Liou, Y.-A., Hsu, P.-C., Lee, Y.-C., Wang, C.-H., & Chao, T.-K. (2022). Histopathological whole slide image dataset for classification of treatment effectiveness to ovarian cancer. *Scientific Data*, *9*(1), 25. DOI: 10.1038/s41597-022-01127-6 PMID: 35087101

Wang, X., Liu, L., & Zhang, Y. (2015). A novel chaotic block image encryption algorithm based on dynamic random growth technique. *Optics and Lasers in Engineering*, *66*, 10–18. DOI: 10.1016/j.optlaseng.2014.08.005

Wang, X., Teng, L., & Qin, X. (2012). A novel colour image encryption algorithm based on chaos. *Signal Processing*, *92*(4), 1101–1108. DOI: 10.1016/j.sigpro.2011.10.023

Wang, X., Xue, G., Huang, S., & Liu, Y. (2023). Underwater object detection algorithm based on adding channel and spatial fusion attention mechanism. *Journal of Marine Science and Engineering*, *11*(1116), 1116. Advance online publication. DOI: 10.3390/jmse11061116

Wang, Y. (2021). *An Automatic Knee Osteoarthritis Diagnosis Method Based on Deep Learning: Data from the Osteoarthritis Initiative*. Hindawi.

Wang, Y. L., Gao, S., Xiao, Q., Li, C., Grzegorzek, M., Zhang, Y. Y., Li, X. H., Kang, Y., Liu, F. H., Huang, D. H., Gong, T. T., & Wu, Q. J. (2024, March 11). Role of artificial intelligence in digital pathology for gynecological cancers. *Computational and Structural Biotechnology Journal*, *24*, 205–212. DOI: 10.1016/j.csbj.2024.03.007 PMID: 38510535

Weber, R., Mangus, J. M., & Huskey, R. (2015). Brain Imaging in Communication Research: A Practical Guide to Understanding and Evaluating fMRI Studies. *Communication Methods and Measures*, *9*(1-2), 5–29. DOI: 10.1080/19312458.2014.999754

Wei, J., Wang, S., & Huang, Q. (2020). $F^3$Net: Fusion, feedback and focus for salient object detection. *Proceedings of the 2020 AAAI Conference on Artificial Intelligence (AAAI)*, 34(7), 12315-12322. DOI: 10.1609/aaai.v34i07.6916

Wei, S., Zhang, Z., & Kim, M. (2018). Secure and Efficient Inference Engine with Deep Learning. *2018 IEEE International Symposium on Multimedia (ISM)*, 455-459. [DOI: ]DOI: 10.1109/ISM.2018.00092

Woon, S. G. (2020). Digital image enhancement. In *Digital Image Processing*. Springer., DOI: 10.1007/978-981-10-5550-8_12

Wu, Z., Xiong, Y., Yu, S. X., & Lin, D. (2018). Unsupervised Feature Learning via Non-parametric Instance Discrimination. *2018 IEEE/CVF Conference on Computer Vision and Pattern Recognition*, 3733-3742. DOI: 10.1109/CVPR.2018.00393

Wu, Z., Xiong, Y., Yu, S. X., & Lin, D. (2018). Unsupervised feature learning via non-parametric instance discrimination. In *Proceedings of the IEEE conference on computer vision and pattern recognition* (pp. 3733-3742).

Xia, Z., Wang, X., Wang, C., Wang, C., Ma, B., Li, Q., & Zhao, T. (2022). A robust zero-watermarking algorithm for lossless copyright protection of medical images. *Applied Intelligence*, *52*(1), 607–621. DOI: 10.1007/s10489-021-02476-2

Xu, W., Liu, H., Wang, X., & Qian, Y. (2019, July). Liver segmentation in CT based on ResUNet with 3D probabilistic and geometric post process. In *2019 IEEE 4th International Conference on Signal and Image Processing (ICSIP)* (pp. 685-689). IEEE.

Xu, Y., Hosny, A., Zeleznik, R., Parmar, C., Coroller, T., Franco, I., Mak, R. H., & Aerts, H. J. W. L. (2019, June 1). Deep Learning Predicts Lung Cancer Treatment Response from Serial Medical Imaging. *Clinical Cancer Research*, *25*(11), 3266–3275. DOI: 10.1158/1078-0432.CCR-18-2495 PMID: 31010833

Yang, Q., Yan, P., Zhang, Y., Yu, H., Shi, Y., Mou, X., Kalra, M. K., Zhang, Y., Sun, L., & Wang, G. (2018). Low-dose CT image denoising using a generative adversarial network with Wasserstein distance and perceptual loss. *IEEE Transactions on Medical Imaging*, *37*(6), 1348–1357. DOI: 10.1109/TMI.2018.2827462 PMID: 29870364

Yang, Y., Liang, W., Zhou, D., Zhang, Y., & Xu, G. (2023). Object detection for underwater cultural artifacts based on deep aggregation network with deformation convolution. *Journal of Marine Science and Engineering*, *11*(2228), 2228. Advance online publication. DOI: 10.3390/jmse11122228

Yao, J., Zhu, X., Jonnagaddala, J., Hawkins, N., & Huang, J. (2020). Whole slide images based cancer survival prediction using attention guided deep multiple instance learning networks. *Medical Image Analysis*, *65*, 101789.

Yarberry, W. (2021). *Regular Expressions: Introduction. CRAN Recipes*. Springer.

Yi, X., Yu, S., & Bertino, E. (2018). Secure Multi-Party Deep Learning Computation using Secret Sharing and GPUs. *2018 IEEE International Conference on Parallel & Distributed Processing with Applications (ISPA)*, 677-686. [DOI: ]DOI: 10.1109/ISPA.2018.00115

Yosinski, J., Clune, J., Bengio, Y., & Lipson, H. (2014). How transferable are features in deep neural networks? *Advances in Neural Information Processing Systems*, •••, 27.

Zebari, D. A. (2022). Knee Osteoarthritis Detection Using Deep Feature Based on Convolutional Neural Network. *International Conference on Computer Science and Software Engineering (CSASE)* (pp. 259-264). IEEE. DOI: 10.1109/CSASE51777.2022.9759799

Zelisse, H. S., Hwan, R. A., van de Vijver, M. J., Dijk, F., Mom, C. H., Hooijer, G. K. J., van Gent, M. D. J. M., & Snijders, M. L. H. (2024). Evaluation of the prognostic potential of histopathological subtyping in high-grade serous ovarian carcinoma. *Virchows Archiv*. Advance online publication. DOI: 10.1007/s00428-024-03807-7 PMID: 38622323

Zeng, C., Liu, J., Li, J., Cheng, J., Zhou, J., Nawaz, S. A., Xiao, X., & Bhatti, U. A. (2024). Multi-watermarking algorithm for medical image based on KAZE-DCT. *Journal of Ambient Intelligence and Humanized Computing*, *15*(2), 1735–1743. DOI: 10.1007/s12652-021-03539-5

Zermi, N., Khaldi, A., Kafi, M. R., Kahlessenane, F., & Euschi, S. (2022). An SVD values ordering scheme for medical image watermarking. *Cybernetics and Systems*, *53*(3), 282–297. DOI: 10.1080/01969722.2021.1983700

Zhang, L., & Xu, L. (2018, June). An automatic liver segmentation algorithm for CT images U-net with separated paths of feature extraction. In *2018 IEEE 3rd International Conference on Image, Vision and Computing (ICIVC)* (pp. 294-298). IEEE. DOI: 10.1109/ICIVC.2018.8492721

Zhang, D., He, Z., Zhang, X., Wang, Z., Ge, W., Shi, T., & Lin, Y. (2023). Underwater image enhancement via multi-scale fusion and adaptive color-gamma correction in low-light conditions. *Engineering Applications of Artificial Intelligence*, *126*, 106972. Advance online publication. DOI: 10.1016/j.engappai.2023.106972

Zhang, G., Tzoumas, S., Cheng, K., Liu, F., Liu, J., Luo, J., Bai, J., & Xing, L. (2017). Generalized adaptive Gaussian Markov random field for x-ray luminescence computed tomography. *IEEE Transactions on Biomedical Engineering*, *65*(9), 2130–2133. DOI: 10.1109/TBME.2017.2785364 PMID: 29989945

Zhang, W., Li, G., & Ying, Z. (2017). A new underwater image enhancing method via color correction and illumination adjustment. *Proceedings of the 2017 IEEE Visual Communications and Image Processing (VCIP)*, 1-4. DOI: 10.1109/VCIP.2017.8305027

Zhang, X., Tian, J., Deng, K., Wu, Y., & Li, X. (2010). Automatic liver segmentation using a statistical shape model with optimal surface detection. *IEEE Transactions on Biomedical Engineering*, *57*(10), 2622–2626. DOI: 10.1109/TBME.2010.2056369 PMID: 20615804

Zhao, C. (2019) Non-parametric Model for Background Subtraction. In: *Proceedings of the European conference on computer vision*, pp 751–767

Zhao, C., Sain, A., Qu, Y., Ge, Y., & Hu, H. (2019). Background subtraction based on integration of alternative cues in freely moving camera. *IEEE Transactions on Circuits and Systems for Video Technology*, *29*(7), 1933–1945. DOI: 10.1109/TCSVT.2018.2854273

Zikic, I., Ioannou, Y., & Criminisi, A.. (2014). Segmentation of brain tumor tissues with convolutional neural networks. *Proceedings of BRATS-MICCAI.*

Zuiderveld, K. (1994). Contrast limited adaptive histogram equalization. In Heckbert, P. S. (Ed.), *Graphics Gems IV* (pp. 474–485). Academic Press., DOI: 10.1016/B978-0-12-336156-1.50061-6

# About the Contributors

**B. Omkar Lakshmi Jagan** present working as Associate Professor in Department of Computer Science and Engineering, Vignan's Institute of Information Technology (A), Duvvada, Visakhapatnam, Andhra Pradesh, India. He worked as Assistant Professor in Artificial Intelligence and Machine Learning Department, School of Engineering, Malla Reddy University, Maisammaguda, Dulapally, Hyderabad, Telangana. He received his B.Tech. degree from Sri Sivani College of Engineering, affiliated to JNTU Kakinada, A P, India in 2014, and his M.Tech. degree from the Koneru Lakshmaiah Education Foundation (Deemed to be University), Vaddeswaram, Guntur, Andhra Pradesh, India in 2016 and also he completed his Ph.D. from the Koneru Lakshmaiah Education Foundation (Deemed to be University), Vaddeswaram, Guntur, Andhra Pradesh, India in 2023 under the Electrical and Electronic Engineering Department. He published several papers in International Journals under the field of Statistical Signal Processing, Biomedical Signal Processing and Image Processing technologies. He is a member of IAENG, IEEE and OSI. He is also a senior member for ISRD and SDIWC. Email Id: omkarjagan@yahoo.com

**Amrit Mukherjee** (Senior Member, IEEE) received the Ph.D. degree in electronics engineering from the Kalinga Institute of Industrial Technology University, Bhubaneswar, India, in 2017.,From 2018–2020, he was a Postdoctoral Fellow with Jiangsu University, Zhenjiang, China, and a Project Engineer in C-DAC (Armenia and Ghana). He is currently with the Department of Computer Science, Faculty of Science, University of South Bohemia, Ceská, Czechia. He has authored or coauthored more than 60 research articles, one patent, and filed four patents. His research interests include artificial intelligence, next-generation IoT systems, wireless sensor networks, cognitive radio, and signal processing.,Dr. Mukherjee was a Guest Editor for several special issues of Computer Communications, Multimedia

Tool and Applications, Journal of System Architecture, Computers, IEEE Internet of Things Journal, etc.

**Thayyaba Khatoon Mohammed** is a Professor & Dean in the Department of CSE-AIML at Malla Reddy University. She completed her Ph.D. and M.Tech. from JNTUH and completed her B.Tech. from Kalinga University. She has 18 years of experience in teaching both UG and PG and research experience of 13 years. She also has 13 years of experience as Head of the Department. She has been guiding 6 Ph.D scholars where 2 scholars have cleared their Ph.D.'s. Her area of expertise is Artificial Intelligence, Nature Language Processing, Network Security and Cloud Computing. She has published several papers in reputed journals and conferences. She has also received young scientist award from IARDO and received funds from DST. Dr.Thayyaba Khatoon Mohammad is a passionate teacher and researcher. She is always willing to share her knowledge with others. She is also a regular speaker at FDPs and workshops. She is looking forward to sharing her knowledge and experience with the participants of this FDP.

***

**Naveena Tresa Joseph** is a dedicated and passionate research scholar currently doing research in computer vision and deep learning, particularly in the context of underwater imagery. Her research path unfolds at APJ Abdul Kalam Technological University (APJKTU) at Amal Jyothi College of Engineering, where she is committed to exploring the enigmatic depths of the underwater realm using state-of-the-art technological innovations. She has a background in computer science and a passion for image processing and her research journey is with a vision to bridge the gap between technology and the oceanic environment.

**Chandan Kumar** is currently working as Associate Professor in computer science and engineering department at Career Point University Hamirpur(H.P.). His research interests include data hiding techniques, image processing, and computer networks. Contact him at chandansharmahmr@gmail.com

**A.V.Senthil Kumar** to his credit he has industrial experience for five years and teaching experience of 27 years. He has also received his Doctor of Science (D.Sc in Computer Science). He has to his credit 33 Book Chapters, 220 papers in International and National Journals 60 papers in International Conferences in International and National Conferences, and edited 12 books (IGI Global, USA). He is as Associate Editor of IEEE Access. He is an Editor-in-Chief for many journals and Key Member for India, Machine Intelligence Research Lab (MIR Labs). He is an

Editorial Board Member and Reviewer for various International Journals. He is also a Committee member for various International Conferences. He is a Life member of International Association of Engineers (IAENG), Systems Society of India (SSI), member of The Indian Science Congress Association, member of Internet Society (ISOC), International Association of Computer Science and Information Technology (IACSIT), Indian Association for Research in Computing Science (IARCS), and committee member for various International Conferences.

**Binod Kumar** is Professor and Dean(International Relations) at JSPM's Rajarshi Shahu College of Engineering (MCA Dept.), Pune, affiliated with Savitribai Phule Pune University, India. He received the IEEE Pune Section Senior Member of the Year 2023 Award. He received Dignitary Fellow of International Organization for Academic and Scientific Development (IOASD) in 2021. He have been nominated for "Distinguished Professor Award" from the Bestow Edutrex International Award 2020. He worked as an associate professor at the School of Engineering and Computer Technology, Quest International University, MALAYSIA .He published 12 patents and two under process. Under his supervision 07 students have completed PhD and 06 students are pursuing PhD at Savitribai Phule Pune University, Pune, India. He has conducted PhD viva-voice as External Expert of 13 students at different universities.He is reviewer of Journals like Elsevier, Springer and TPC of various IEEE sponsored conferences. He is Editorial Board member of 45 International Journals. Editorial Review Board of Journal of Comparative Asian Development (JCAD), IGI Global USA.

**S. N Kumar** is currently an Associate. Professor, Dept. of Electrical and Electronics Engineering., Amal Jyothi College of Engineering, Kanjirappally, Kerala. He was awarded PhD in Electronics and Communication Engineering from Sathyabama Institute of Science and Technology, Chennai. He has 15 years of teaching experience, and his areas of interest include Medical Image Processing and Embedded System Applications in Telemedicine. He was the Co- Principal Investigator of the DST IDP-funded project and research fellow of the RCA scheme, NTU Singapore. He has edited 3 books, authored 32 book chapters and 30 publications in peer reviewed journals. He had authored 12 text books in engineering discipline.

**R. Parvathi** is a Professor of School of Computing Science and Engineering at VIT University, Chennai since 2011. She received the Doctoral degree in the field of spatial data mining in the same year. Her teaching experience in the area of computer science includes more than two decades and her research interests include data mining, big data and computational biology.

**Binit Roy**, a researcher in ICFAI University's Computer Science Engineering Department, holds an M.Tech from Tripura Central University and a B.Tech from West Bengal University of Technology. With a year of experience as a visiting faculty at ICFAI, Binit specializes in data science, machine learning, and artificial intelligence. His solid academic background and passion for computer science fuel his research in innovative problem-solving. As a research scholar, he collaborates with peers, publishes in top journals, and constantly strives for excellence, making him a valuable contributor to the field of computer science engineering.

**C.V. Suresh Babu** is a pioneer in content development. A true entrepreneur, he founded Anniyappa Publications, a company that is highly active in publishing books related to Computer Science and Management. Dr. C.V. Suresh Babu has also ventured into SB Institute, a center for knowledge transfer. He holds a Ph.D. in Engineering Education from the National Institute of Technical Teachers Training & Research in Chennai, along with seven master's degrees in various disciplines such as Engineering, Computer Applications, Management, Commerce, Economics, Psychology, Law, and Education. Additionally, he has UGC-NET/SET qualifications in the fields of Computer Science, Management, Commerce, and Education. Currently, Dr. C.V. Suresh Babu is a Professor in the Department of Information Technology at the School of Computing Science, Hindustan Institute of Technology and Science (Hindustan University) in Padur, Chennai, Tamil Nadu, India. For more information, you can visit his personal blog at https://sites.google .com/view/cvsureshbabu/.

**Kadir Uludag** holds Ph.D. in Applied psychology (Chinese Academy of Sciences). He is currently doing postdoc in Shanghai Jiaotong University Mental Health Center. His research interest includes schizophrenia research, drug addiction and educational psychology. In addition, he runs a website to share and comment on peer-reviewed articles (ifx0.com). I am enthusiastic about collaborative opportunities that allow me to contribute to society. Please don't hesitate to reach out to me if you'd like to engage in any collaborative endeavors.

**Xiaohui Yuan** received the B.S. degree in electrical engineering from the Hefei University of Technology in 1996 and the Ph.D. degree in computer science from Tulane University in 2004. He was with the National Institute of Health between 2004 and 2006 and is currently a Professor with the University of North Texas. His research interests include computer vision, artificial intelligence, data mining, and machine learning. He was a recipient of the Ralph E. Powe Junior Faculty Enhancement Award in 2008. He serves as the Editor-in-Chief of the International

Journal of Smart Sensor Technologies and Applications and on the Editorial Board of several international journals as well as the Chair of many international conferences.

# Index

## A



## B



## C



## D



## E



## G



## H



## I

## L



## M



## N



## P



## R



## S



## U

www.ingramcontent.com/pod-product-compliance
Ingram Content Group UK Ltd.
Pitfield, Milton Keynes, MK11 3LW, UK
UKHW050947290926
15470UKWH00005B/1694

9 798369 338407